NORTH
TO THE FUTURE

NORTH
TO THE FUTURE

AN OFFLINE
ADVENTURE THROUGH
THE CHANGING WILDS
OF ALASKA

BEN WEISSENBACH

GRAND CENTRAL

New York Boston

Copyright © 2025 by Ben Weissenbach
Jacket design and illustration by Jim Tierney.
Jacket copyright © 2025 by Hachette Book Group, Inc.

Hachette Book Group supports the right to free expression and the value of copyright. The purpose of copyright is to encourage writers and artists to produce the creative works that enrich our culture.

The scanning, uploading, and distribution of this book without permission is a theft of the author's intellectual property. If you would like permission to use material from the book (other than for review purposes), please contact permissions@hbgusa.com. Thank you for your support of the author's rights.

Grand Central Publishing
Hachette Book Group
1290 Avenue of the Americas, New York, NY 10104
grandcentralpublishing.com
@grandcentralpub

First Edition: July 2025

Grand Central Publishing is a division of Hachette Book Group, Inc. The Grand Central Publishing name and logo is a registered trademark of Hachette Book Group, Inc.

The publisher is not responsible for websites (or their content) that are not owned by the publisher.

The Hachette Speakers Bureau provides a wide range of authors for speaking events. To find out more, go to hachettespeakersbureau.com or email HachetteSpeakers@hbgusa.com.

Grand Central Publishing books may be purchased in bulk for business, educational, or promotional use. For information, please contact your local bookseller or the Hachette Book Group Special Markets Department at special.markets@hbgusa.com.

Map by Julia Ditto

Print book interior design by Jeff Stiefel

Library of Congress Cataloging-in-Publication Data

Names: Weissenbach, Ben author
Title: North to the future : an offline adventure through the changing wilds of Alaska / Ben Weissenbach.
Description: First edition. | New York, NY : GCP, 2025.
Identifiers: LCCN 2025001088 | ISBN 9781538758335 hardcover | ISBN 9781538758359 ebook
Subjects: LCSH: Weissenbach, Ben | Adventure and adventurer—United States—Biography | Alaska—Description and travel | Climatic changes—Alaska | Landscapes—Alaska | Nature—Effect of human beings on—Alaska | LCGFT: Autobiographies
Classification: LCC CT9971.W38 A3 2025 | DDC 979.8/052092 $a B—dc23/eng/20250402
LC record available at https://lccn.loc.gov/2025001088

ISBNs: 9781538758335 (hardcover), 9781538758359 (ebook)

Printed in the United States of America

LSC-C

Printing 1, 2025

For John McPhee

North to the Future
　　　—Alaska's state motto, adopted in 1967

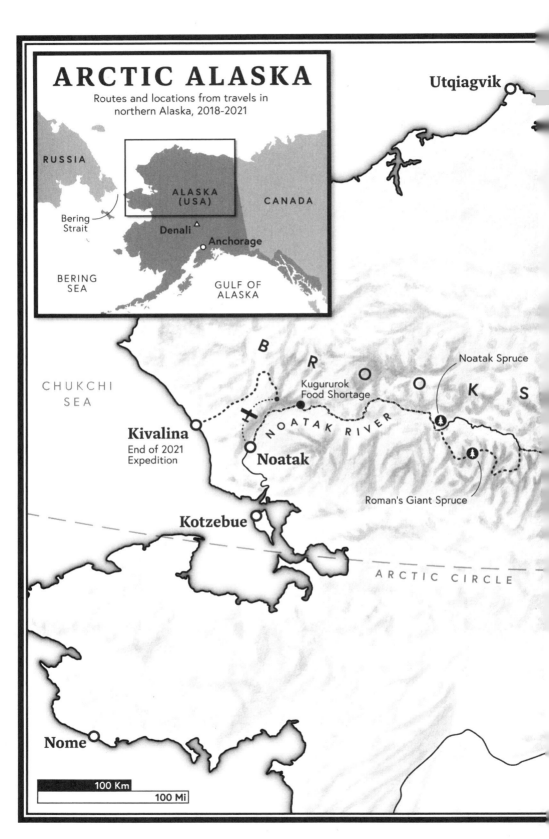

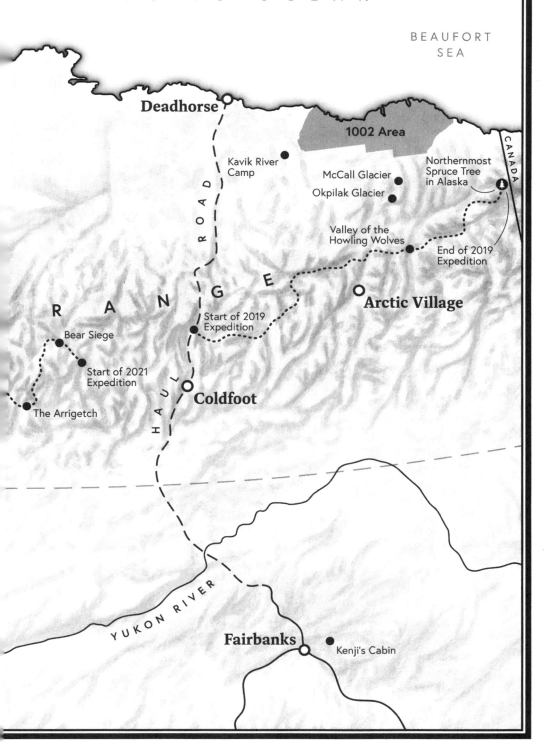

CONTENTS

Prologue – 1

Part I: The Migrating Forest – 5

Part II: Shifting Foundations – 81

Part III: Maps of McCall Glacier – 149

Part IV: North to the Future – 195

Epilogue – 297

Acknowledgments – 303

NORTH
TO THE FUTURE

PROLOGUE

June 4, 2021

Down below the bear is digging. With one paw he parts the thin, wind-compacted soil, combing the tundra for edible roots and sweet overwintered berries. His movements are slow, languorous, as though he is in no rush to get at us. Like the tiresome work of cornering us is done, and we are now livestock trapped in his sprawling corral. Or maybe this is what a desperate bear looks like; maybe he is gathering his energy, scrounging a few calories before the final struggle.

I can't tell for certain. I have never before had so much time to watch a grizzly bear's movements—have never before seen how its sun-bleached fur quivers and vibrates in the light, as if each hair were alive. This is what the traditional Koyukon Athabaskans, who hunted in these mountains for centuries, have long believed: Each hair on the grizzly is alive. They used to wait several years after killing a bear to use its hide. That's how long it took for all that life to leave the body.

Beside me on this rocky outcrop, 150 yards above the bear, Roman is napping. Roman Dial: ecologist, explorer, and arguably the most skilled outdoorsman in Alaska. He lies on his yellow sleeping pad, the wind rippling his white hair. Next to him, my tentmate Julia sits listening to the swallows. We are tired. It has been only thirty-two hours since the pilot deposited us in these mountains—the Brooks Range of Arctic Alaska, among the northernmost mountains in the world—but the bear has been following us for at least twenty-six of them. "Couldn't he have waited until, like, day five to come for us?" whispered Russell, Roman's graduate student, two hours ago. We were hiding in a thicket of alders, and Roman shushed him.

In the hours since, we have gathered rocks to throw at the bear. We've talked through escape strategies. If we can get past him and continue our journey west, we will have two and a half months to reach the Arctic Ocean. Two and a half months to walk and raft six hundred miles through the fastest-warming region on the planet, which also happens to be a roadless wilderness the size of California. We've come to study a new forest that only a handful of people know exists and which Roman believes may be the fastest-migrating forest in the world. To peer into the future of Northern Alaska, which is also, in a way, the future of life on Earth.

Things are changing here in this land of all-day darkness and all-night light. Things have always changed here. You can see it in the ghastly fossils that fall out of riverbanks, in the stories Native peoples tell, and in the smooth flanks of mountains, carved down by waves of ice that come and go over thousands of years. But this planet now seems to be warming faster than at any time since, at least, the invention of the written word, and the Arctic is warming about three times faster than the rest. Already this is triggering a cascade of ecological and biogeochemical changes, which computer models say might radically compound warming the world over. The greatest remaining uncertainties, though—and the uncertainties remain great—cannot be resolved by computing power alone. Nor can the models tell us what these changes might mean for us, our children, and their children's children. To approach this mystery, one of the greatest of our age, takes time, context, and careful attention.

It is hard and rare, especially these days, to pay this kind of attention. The realm our screens bring close is so captivating, and so damn *useful*, that it can be hard to put down for long enough to take a good look at the world itself. When I first arrived in Alaska three years ago, a twenty-year-old from Los Angeles, I didn't even know how. I could walk for miles and miles through a strange land without ever encountering its strangeness, without ever leaving the nervous chatter of my own mind. I took pictures and notes constantly; very little got through.

It has taken many months in this place—a previous trek with Roman across the eastern Brooks Range, a winter stay with permafrost scientist Kenji Yoshikawa at his off-grid cabin in the interior, and a trip with glaciologist Matt Nolan to the highest mountains in Arctic Alaska—to begin to learn how to pay this world its due. To begin to learn the patterns of its plants and animals, the changing rhythms of its weather and seasons, and the way time blows through it all, making a beautiful joke of our theories. I am still learning. But I have found that the story of change, here on the frontier of change, is far more complex than I could have imagined. And I am finding, too, that I am in love with this world.

The greatest mysteries have a strange way of clarifying things. We are all scared, watching the bear, but this is life for now. The escape strategies have been gamed out. The throwing stones have been gathered. So Roman naps, and Julia listens to the swallows, and Russell studies the map, while I weigh the past three years like the rocks in my hand.

PART I

THE MIGRATING FOREST

> The attraction of knowledge would be small if one did not have to overcome so much shame on the way.
> —Friedrich Nietzsche, *Beyond Good and Evil*

> And you may ask yourself, "Well, how did I get here?"
> —Talking Heads, "Once in a Lifetime"

1

I met Roman near the end of my first summer in Alaska, the summer of 2018. I was a rising junior in college then and had received a university grant to pursue a journalism project, which was a thinly veiled excuse to spend three months traveling across a region that had long captured my imagination, asking people, "If you could talk to one person in all of Alaska, who would it be?" Then I would try to find that person, interview them, and ask the same question.

In a state of 730,000, I began to hear the same names again and again, and many of them belonged to ecologists, biologists, glaciologists—people studying change in the land. That year was the fifth of five straight that surpassed the previous global temperature record, and Alaska was warming particularly fast, even for the Far North. Locals told me of dead seabirds lining their coasts, strange fungi in the water, and treacherous ice conditions. Commercial fishermen spoke of unprecedented fluctuations in their salmon harvests, hunters mentioned shifting migration patterns, and skiers lamented utterly unshreddable snowpack. Since most of Alaska is impenetrable to most scientists, a relatively small and hardy group had emerged as key figures in a kind of scientific Wild West. Roman Dial was one of these figures—along with Matt Nolan and Kenji Yoshikawa, of whom I will say more later—but I was too intimidated to contact him directly. We might never have met but for a quirk of fate.

As it happened, on the evening of August 17, about a week before I was scheduled to fly home, I left my low-rent hostel in Anchorage and Ubered to a party on the far side of town. The party was for a prominent outdoorsman named Luc Mehl, whom I had met, after sending a cold email, several weeks before; tonight he was turning forty, and he'd generously invited me to drop by.

I arrived just after eight, clutching a six-pack of IPAs, and found a laid-back gathering on the patch of lawn behind his house—a few dozen people pulling from kegs in the sweetening end-of-summer light. Among the guests were some of the most accomplished outdoorspeople in the state: men and women who slipped off into the mountains to scale previously unclimbed peaks and raft big, remote rivers, and then, in typical Alaskan fashion, came home and told almost no one about it.

Soon after I arrived, the chatting turned to laughter and cheers as two men tore across the lawn, hurtling through what appeared to be an obstacle course—or, as I would later hear Luc call it, an "adventure course." The name referred to an "adventure race" called the Alaska Mountain Wilderness Classic, which every summer and winter sees between ten and thirty participants race across more than a hundred miles of unmarked terrain more rugged, wild, and unforgiving than can be found in the Lower 48. The event is little known outside of Alaska, mostly due to efforts to avoid attracting reckless yahoos, since it is almost certainly one of the more grueling events in the modern world. At the 2012 Winter Classic, for example, Luc and a friend skied two hundred miles through Arctic mountains in just under four days, enduring −20°F temperatures and bouts of hallucination.

Luc had won several Classics in recent years, but arguably the most distinguished competitor in the event's history was Roman Dial. He won the inaugural Classic in 1982, and was still dominant enough in 2002—on a 150-mile course that passed through the heart of the heavily glaciated Wrangell Mountains, and demanded slogging through a blizzard and packrafting Class III rapids—to help save a hypothermic competitor while setting a course record of two days and four hours. Due to a rash, he ran into the finish wearing nothing but a makeshift loincloth.

"One common trait of the race is that just about everyone suffers to a certain degree with mild hypothermia," Roman once told a journalist. A former winner and Vietnam veteran was reported to have called the race "as close as a civilian can come to experiencing actual combat." The release statement is explicitly discouraging; in the eighties, when Roman was race director, it began: "IT AIN'T OUR PROBLEM AND YOU CAN'T SUE US FOR ANYTHING." In 2014, a veteran participant drowned.

And yet the Classic is more than a thrill-seeker's proving ground. Alaska remains, well into the twenty-first century, forbiddingly remote, and only those with a deep practical knowledge of the land can approach its mysteries. In the contiguous US, there is no point more than twenty-one miles as the crow flies from the nearest road; if you ever get lost, you can pick a random point on the horizon and walk toward it, and within a day or three you'll be sure to hit human infrastructure. But Alaska is bigger than California, Texas, and Montana combined, and has swaths of wilderness that could swallow Eastern states whole and burp out Rhode Island. If a bush pilot were to drop you into Alaska at random, you would be lucky to land within twenty-one miles of one of the state's four highways. You could, to borrow a Roman-ism, "stumblefuck" for weeks through icy rivers, big mountains, and thick brush, and never see sign of another human.

The Wilderness Classic, then, is less a sporting event than a modern celebration of the ancient art of reading the land. The race offers no prize. It doesn't even call itself a race, for that matter. The professed goal is to compete with yourself, to test and deepen your knowledge of the bush. The best competitors learn to do without tents, stoves, or even sleeping bags—that is, to shave their margin of error down to almost zero. Leaving this equipment behind requires knowing, among other things, how to read the skies for inclement weather, how to start a fire quickly, how to ford big rivers, how to navigate on the fly. It demands (aside from a touch of masochism) a capacity to open one's senses to the land and a willingness to encounter it on something like its own terms.

Not that I really understood this at the time, and not that Luc's birthday "adventure course" called for any of it: Racers bobbed for apples floating alongside red fuel canisters, threw darts at a map of Alaska, and ran dizzying laps around the inside of an old dome tent. The obstacles had names like Pin the Leuko on the Blister,[1] Bushwack Balance, and Bear Mace to the Face. (This last challenge was not so brutish as it sounds; participants simply had to spray Silly String at a friend of Mehl's dressed in a dog costume as he popped out from behind a willow-adorned tarp.) I was standing next to Topo Tire Toss—a challenge in which racers ripped a map out of an atlas, folded it into an airplane, and lofted it through a bike tire—when I spotted Roman talking to some people by the kegs.

I had, for the most part, been dogged about meeting people that summer, but I'd hesitated to contact Roman. People spoke of him as a modern wizard of the wilderness, a cat with nine lives, a guy who should,

---

1 *Leuko* referred to Leukotape, a brand of athletic tape often used to treat blisters.

by all rights, have died a long time ago. When he was twenty-four, for example, he and a climbing partner made a first ascent (in winter, no less) of a hellish chunk of the Alaska Range called Cutthroat Couloir. While descending along a knife-edge ridge, his partner broke through an overhung cornice and began to free-fall. As the rope that connected them slithered through Roman's mittens, threatening to pull him off the ridge in his partner's tow, he chose, in a moment of quick thinking, to launch himself off the opposite side of the ridge. Roman fell two hundred feet before the rope jerked him, and his injured partner on the other side, to a dangling halt.

Now fifty-seven, Roman had turned his full attention to science and was a professor of math and biology at Alaska Pacific University. Having crossed the line between risky and reckless enough times to learn exactly where it was, he'd made a career of operating in environments where few others could. Research had taken him to Alaskan ice fields to study snow algae, to Himalayan glaciers in search of ice worms, and to the tops of two-hundred-foot-tall *Koompassia excelsa* trees in Borneo to study forest canopies. (There he used a pistol-grip crossbow to shoot ropes over branches, shimmying horizontally from tree to tree.) And though I couldn't have known it then, he was entering his golden era, a run of research that would appear in top journals like *Nature* and *Science*.

I, meanwhile, held the distinction of being part of the first generation to go through adolescence with front-facing cameras. By the time I turned eighteen, I had logged an ungodly number of hours on YouTube, Facebook, and Snapchat, but no more than a handful of nights in a tent. Anxious, sensing that something might have been lost in my cohort's mass migration to the so-called cloud, I'd taken a year between high school and college to learn how to be outside, during which I'd backpacked in South America, hiked off-trail in Northern California, and become informed on things like how to shoot compass bearings and shit in the woods. But I still had the perennially distracted mind of a digital native and the know-how of a cheechako (the Yukon term for greenhorn). The Alaskan wilderness remained to me a remote, albeit powerful, idea.

That idea was really one of escape: a land still mostly beyond modern human history, where one could get away from all the dinging and buzzing. Where people were undistracted, and knew what was real, and did things instead of just talk about them. It was an idea I had selectively drawn from books like Jack London's *The Call of the Wild* and especially John McPhee's 1977 classic *Coming into the Country*, which described back-to-the-land bush-dwellers, ten-thousand-year-old Native cultures, and a country that remained vast, raw, and indifferent to the machinations of modern humankind. I was skeptical that a place like that still existed but had to know. McPhee was eighty-six by the time I encountered his book, but he still taught a sophomore writing seminar at Princeton, where I was a student. As soon as I was eligible I applied for the course, writing him: "I want to follow in your footsteps—literally."

But doing so would require following in the footsteps of someone like Roman, and I knew just enough to realize I was out of my depth. So it was with some trepidation that I found myself shoulder to shoulder with him by an obstacle called Tussock Town, a bed of stuffed animals covered by a tarp to simulate the uneven footing of the tundra. His hair looked longer and whiter than in photos, but I'd recognized his icy eyes from across the yard. About six feet tall, with bushy eyebrows and a prominent nose, he appeared light on his feet even standing still.

"Dr. Dial?" I said, screwing on a smile.

"I like the sound of that," he said, turning to face me.

NORTH TO THE FUTURE

I told him about my summer project—the places I had been, the people I had met. Up close now he seemed gentle and warm. When I mentioned that I'd designed my project with McPhee's guidance, Roman raised his bushy eyebrows. "His was the best book ever written about Alaska," he said. "Can you walk?"

I looked down to make sure my legs were still there. Too nervous to crack a joke, I told him about a month-long expedition I'd done in the Alaska Range earlier that summer—an ascent of the north side of Denali, North America's highest peak.

He nodded. "What are you doing next weekend?"

"I'll be around," I lied. I had promised my parents that I would return to LA that Friday, in time for my extended family's visit. But my flight could be postponed; getting to know Roman Dial couldn't be.

"I'm going moose hunting," he said. "I could use some help carrying out the meat."

Seven days later, we were nudging rafts into the Tanana River, straining our senses for signs of a cow moose. Quickly I realized that when Roman had asked if I could "walk," he'd meant off-trail: through forests, over mountains, around wild animals. "*Walk more quietly!*" he whispered gravely once, as we followed the fresh tracks of a moose. Over the course of four days spent bushwhacking and rafting and talking around the fire, I began to sense that I had a great deal to learn from Roman, about things I didn't even know I didn't know. There was no way for me to recognize it then, but I was entering his life at precisely the right time.

2

kept in loose contact with Roman that fall, and in February 2019, six months after the moose hunt, he sent another invitation. I was in my college dorm room when the email came in, and I suspect that everyone in the building heard my reaction. I am not a loud person, but I shouted. I am not a good dancer, but I danced.

He'd set his sights, he explained, on one of the more consequential mysteries the natural sciences faced: Over the previous five decades, Arctic Alaska had warmed by more than 5°F—roughly the difference in mean annual temperature between the redwood forests of Northern California and the scrubland deserts of Northern Mexico. No one knew how—or when—this would affect the world's largest land biome, the boreal forest (also known as taiga), which covers one-ninth of Earth's terrestrial surface and stores more carbon than all of the world's temperate and tropical forests combined. Would these great forests flourish and migrate northward? Would they wither and burn, spewing as much carbon dioxide into the atmosphere each year as all of the coal-fired power plants in China?

NORTH TO THE FUTURE

To date, scientists had conducted intensive studies at a smattering of points across the Arctic, but the results were varied and didn't amount to a comprehensive picture. Others had tried using satellite imagery to address the question, but the resolution of these images was too coarse to answer the most pressing questions. Like so many climate-related matters, this was a problem of scale, made harder by the remoteness of the terrain. I had always vaguely assumed, growing up in the Google Earth era, that a coordinated army of scientists and robots had thoroughly sounded even the far-flung corners of the planet—but as I would soon learn, there remain forests and mountains that are little better known than the deep sea. In all of Alaska's eastern Brooks Range, for example—a region roughly the size of Maine—there is not a single long-term weather station. Meteorologists call it an "observational desert."

The question demanded a broad-scale field campaign, and Roman was in a unique position to lead it, having walked, rafted, and skied thousands of miles across Arctic Alaska. The National Science Foundation had given him and his colleague Patrick "Paddy" Sullivan a grant, and now he was planning a six-week, 360-mile expedition across the eastern Brooks Range. It was going to be a "very big, full-bodied trip," his email said, and he still needed one more field-worker. "Maybe that could be you?"

Rather than ask questions, I pitched the story to *Smithsonian* magazine, applied for a grant, and booked a flight to Anchorage. A few days after my last spring exam, Roman picked me up from Ted Stevens International, and I slept the next several nights in the basement of his home—a friendly-looking one-story that was, apart from the climbing-rope carpets and geological curios, surprisingly conventional. I spent those whirlwind days packing food, preparing equipment, and taking a firearms safety course Roman had signed me up for, presumably to make sure I wouldn't kill anyone with the twelve-gauge Mossberg we were bringing to defend against bears. At a shooting range outside of town, a mustachioed man in paramilitary attire taught me how to clean, load, and fire a weapon, all the while regaling me with stories of bewildering incompetence, in which I could not help feeling implicated.

It was also during these days that I met the other five members of our expedition, all of whom Roman had cherry-picked from Alaska Pacific University. Among them was Russell Wong, a broad, towering undergrad from Wauwatosa, Wisconsin, who had resumed college in Alaska after dropping out of Lewis and Clark to backpack in the Himalayas. Dark-haired with soft green eyes, he wore a dirty old fishing hat, and since the semester ended he had been living out of his truck. And then there was Julia Ditto, a diminutive, bob-haired painter and environmental science major from Anchorage, whom Roman had selected, in part, to produce illustrations of our findings. She was also "third-generation outdoors," according to Roman, and had grown up skiing and backpacking the Chugach Mountains with her father. While Russell struck me as almost pathologically mellow—across his forearm was tattooed a maze of Devanagari script that translates, literally, to "cool beans"—Julia had air-dried and packed all of her trip meals with type A+ precision; the calligraphic labeling alone looked like it must have taken days. Both were my age, twenty-one, but in time I would come to admire them nearly as much as I did Roman.

On May 29, we loaded all our food and equipment into a truck to begin the 670-mile drive north. "I don't think I've ever taken so much stuff on a trip before," said Roman. "We're moving to the Brooks Range," joked Julia. I kept my mouth shut. To me, it looked like almost nothing.

NORTH TO THE FUTURE

* * *

There is only one road into the American Arctic. Officially named the Dalton Highway, colloquially known as the Haul Road, it is a narrow, mostly unpaved washboard of a thing that begins near Fairbanks and ends 414 miles north, on the Arctic Sea. Along its course there are no lights, no billboards, no convenience stores or rest stops. By the time we crossed the Arctic Circle, the rear window of the truck was caked dark with layer upon layer of dry, powdery earth.

Running alongside the road was its raison d'être: the Trans-Alaska Pipeline System, a forty-eight-inch-diameter oil pipe jacked up on stilts, like an endless steel centipede. A monument to modern Alaska's conflicted identity. It was the discovery of oil on the Kenai Peninsula in 1957, and the attendant promise of taxable economic activity, that spurred the territory to statehood in 1959. It was the later discovery of far vaster quantities at Prudhoe Bay, on the northern edge of the continent, that triggered the influx of investment required to construct highways, airports, and schools in one of the more expensive places on Earth to build. In this climate of oil-fueled optimism, the state chose its motto: "North to the Future."

In the half century since, the Prudhoe Bay reserves have largely dried up, but the pipeline remains the backbone of Alaska's economy, providing roughly a third of the state's gross domestic product. Meanwhile the atmospheric warming partially caused by the burning of the pipeline's contents is disproportionately affecting Alaska—because greenhouse gases do not merely trap heat; they also give it more time to diffuse from warm regions to cold ones. (This largely explains why Mercury, which has almost no atmosphere, is roughly 1,000°F colder at its poles than its equator, while Venus—which is smothered in carbon dioxide—is all virtually the same balmy temperature.) Alaska's warming has even begun to threaten the pipeline's structural integrity.

A few years ago, engineers noticed that one of the hills the pipeline crossed was beginning to collapse. Frozen ground had previously held the land together, and as it thawed, the ground itself was giving way.

"Our state is threatened," then Lieutenant Governor Byron Mallott had told me during my first summer in Alaska, in the four-thousand-person seaside village of Utqiagvik, where the disappearance of summer sea ice was causing rapid coastal erosion. "We need to be transitioning to renewable energy," he continued. But oil is what pays for the Alaska Permanent Fund, which distributes a de facto reverse income tax to residents, providing the supplemental cash many village dwellers rely upon to subsidize their semi-subsistence lifestyles.[1] Oil and gas revenues also provide the lion's share of most villages' budgets, including Utqiagvik's. What Mallott said next captured the contradiction at the heart of twenty-first-century Alaska: "We will be looking for streams of revenue that will help us transition to renewable energy. We support continuing petroleum development as long as there's a market."

Roman was no fan of Alaska's oil dependence, but the road would make our logistics easier. Marginally easier. The eastern Brooks Range is one of the few places on Earth where one can travel for weeks at a time without encountering a trace of humankind, aside from the occasional jet contrail hanging in the atmosphere. For millennia, a nomadic, Athabaskan-speaking people called the Gwich'in hunted throughout much of the region, but for almost a century now, they have been settled on the south side of the Range. There exists in the region a number of unmodified gravel bars and flat patches of grass that a handful of local bush pilots generously call "airstrips," but when I made the mistake of asking why we weren't simply flying from one to the next, Roman guffawed. "Have you done much bush flying? Have you done *any* bush flying?"

---

[1] Over the past decade, the state has paid out nearly $10 billion in cash to residents. The permanent fund dividend (PFD) averaged $1,580 per recipient, per year.

NORTH TO THE FUTURE

As he proceeded to explain, taking off in a plane small enough to land on these makeshift strips requires optimal weather—not something you can count on in the Brooks Range. Not to mention the danger inherent to remote bush flying. ("There are old pilots and bold pilots," goes an Alaskan saying, "but there are no old, bold pilots.") A few years before, Roman had been returning from a winter moose hunting trip in a Super Cub (a two-seater with a fabric-covered fuselage) when the plane's single engine cut out. He watched the pilot rap his knuckles on the fuel tank to check its fullness and mutter "Oh, shit." Miraculously, the pilot landed the plane safely on a frozen lake, and they spent a frigid night huddled around a fire before another pilot dropped off gas the next morning. "I'm too old for that sort of thing," said Roman.

So he'd decided, naturally, to *walk* through the mountains and *paddle* the rivers in packrafts—inflatable boats that weigh five to eight pounds each and pack down to the size of a paper towel roll. We'd begin from the road, leveraging Roman's adventure racing experience to trim our camping gear light enough to schlep science equipment for ten to fifteen off-trail miles per day. A bush pilot would deposit barrels of food and science equipment every fifty to eighty miles along our route to sustain us, and after 360 miles and forty-two days, he would pick us up by Alaska's eastern border with Canada.

This mode of travel offered an unquantifiable advantage, which I now suspect motivated Roman's thinking all along: We'd see more. "Walking across the landscape," he explained, "I'm able to develop my intuition about what's going on." But it came with drawbacks, one of which I found particularly sobering: Deep in grizzly country, Roman and I—who were sharing a tent—would have to sleep with the entire group's food. Bear-proof storage barrels were too heavy to haul 360 miles, and we couldn't hang our food from trees, either, for what few trees survived this far north were too short, their branches well within reach for a standing bear. So, in flagrant contradiction of all conventional wisdom—"Store all food away from your campsite," says the Alaska Department of Fish and Game's bear guide, along with every other piece of bear safety literature one might ever wish to read—Roman and I would sleep with the team's food between our bodies and the shotgun behind our heads. "We'll defend our food like a bear would—by sleeping on it," he told me. What "defense" might look like inside a cramped tent with a grizzly, I didn't care to find out.

"I hope this all works," mused Roman as we drove. "It seems pretty improbable, doesn't it?" I nodded woodenly. From the comfort of Princeton, New Jersey, this had seemed like the kind of grand adventure that would be impossible to turn down. But few scientists had traveled this way, this far, since the early twentieth century, and as I watched snowy mountains grow on the horizon, doubts crept in. *What happens if the pilot fails to make one of our food drops? What if one of us snaps a femur a hundred miles from the road?*

We made our last rest stop at a small cluster of buildings—log cabins, a tiny post office, a visitor center—huddled at the southern foot of the Brooks Range. Called Coldfoot, this thirty-four-person outpost is the largest settlement along the Haul Road and the last stop for gas before the Arctic Ocean, 240 miles north. The village got its name in 1900, when a group of gold miners made it all the way to the mountains and then got "cold feet," retreating south to avoid the Arctic winter. As we sat in Coldfoot's only café, eating burgers and fries beside grizzled truckers on their way to and from the oil fields, I realized that I had no real idea what I had signed myself up for. I drank a beer to keep my feet warm.

3

On the evening of June 1, 2019, sixty-two miles north of Coldfoot, we set out east from the road on foot, beginning a five-day journey to our first field site. Years later I recall it vividly: The smooth purple valleys drenched in amber light. The naked huddles of willow and alder shrubs, rubbed raw by wind and ice, and the soft roar of dozens of creeks flushing tons of thawing slurry down to the river below. The whole land groaning, shifting, waking from the long winter. But it's hard to say how much I am now backfilling with more recent impressions; the truth is that in those early days, I was too busy to look around.

There were seven of us at the start, and we all wore enormous packs, lightweight trail sneakers, and soft-shell shirts with the sleeves rolled up. (Temperatures here can, in winter, plummet to −60°F—cold enough to freeze spit before it hits the ground—but from early June to mid-July, the sun does not set and the climate is mild.) Two of the seven, a data technician named Felipe Restrepo and a PhD candidate named Scott Smeltz, were joining for the first two weeks of the trip, before being picked up by the pilot to return home for other work. The remaining three, all undergraduates, would go the full distance, and Roman had vetted each one.

Julia, the painter, had joined him the previous summer on a seventeen-hour day of fieldwork that involved digging a nine-foot-deep hole, skiing across a glacier, trudging through waist-deep meltwater, and biking fourteen miles. "I don't like whiners," she recalled him telling her, and she never whined. Russell already had some outdoor cred from APU's Outdoor Studies department—a program that includes classes such as Mountain Rescue and Backcountry Survival Skills—and had proven his quantitative wits in Roman's notoriously demanding statistics class. The third student, Duncan Wright, an ebullient, burly skier from Anchorage, was asked questions like: "What's the most pain you've ever experienced?"

In my case, the moose hunt the previous summer had apparently sufficed, but it did not take long to establish myself as the expedition's weakest link. On the morning of the first full day, as we nosed our packrafts into a gentle cross current, I flipped clumsily out of my boat and into the icy river. Gasping, I clawed my way to shore, collected myself, tried again, and again swam. The next day, when we encountered our first rapids (still, by Roman's standards, "baby water"), I managed to stay upright, but clenched my jaw so hard that it later ached. I was the only member of the expedition from drought country—the only member, for that matter, who wasn't an Alaska resident—and it showed.

The most demoralizing aspect of these early days, however, was not the river, but Roman's withering gaze. Never had I been in the presence of someone so intentional in his movements or so cuttingly judgmental. On the moose hunt the previous summer, he'd corrected me when I walked too loudly, zipped the tent door too quickly, handled his packraft too brusquely—but he had, on the whole, been gentle. Now, with hundreds of miles before us and NSF money on the line, he was more direct.

"I've learned that most people aren't very careful," he said pointedly one afternoon before explaining how to handle a tape measure that was, like much of his equipment, old and fragile. "It's amazing how much commonsense stuff today people don't know," he jabbed on another occasion. When I misplaced my eating utensil one morning, he remained conspicuously silent, and though I eventually found the thing, camouflaged among river stones, I had already begun to imagine the rest of the trip without it: me grubbing my wretched gruel each night by hand, Roman looking on with open disdain.

The other twenty-one-year-olds felt Roman's scrutiny to varying degrees, too. Duncan, whom Roman once called "a loud Anchorage kid," privately referred to the first week of the trip as "Roman's crucible." But as Roman's only tentmate, I found little respite. Just getting into his tent required presence of mind and body. It was an airy thing, a one-pound sheaf of waterproof fabric that he lofted into a pyramid with two hiking poles lashed together, and to maximize its lifespan, he insisted that we unzip the *top* of the door—far less taut than the bottom—just a couple of feet, and then high-step through the small, eye-shaped hole, like storks wading into a pond. This technique minimized pressure on the tent's zipper, but it also required quite a bit more agility than I was used to needing each time I had to pee—and one windy afternoon, on the eve of our first day of fieldwork, I forgot Roman's instructions: I began to unzip the tent from the bottom.

"I saw that," he barked from across camp.

"It was a mistake," I yelled back, voice fragile with rage and insecurity. "People make them!"

"Yeah," Roman replied. "People also shoot up schools."

Filled with a bottled fury, desperate for his approval, I stewed that night in resentment. But it was a resentment born of shame, because Roman, of course, had a point. In daily life, I was always smushing and jamming, yanking and dropping. Always trying to push through doors carrying too much stuff. I didn't have time to make a second trip or to lower my mug gently enough to avoid sloshing coffee. I didn't have time to thoroughly dry my hands before handling my cracked iPhone.

It's not that I was uninterested in material things. My newsfeeds were filled with well-lit objects I'd never touch: a lathe shaving bars of multicolored soap into hypnotic geometrical patterns, a watermelon blasted to smithereens at a thousand frames per second, a cube of glass compressed by some enormous machine to the point of fracture. Sheer, gross *thing*ness. But I could make it all disappear with the swipe of a finger, and when it came to actually interacting with the world, I seemed to expect the same frictionless compliance.

In Roman's presence this behavior seemed absurd, like I'd spent my whole life running around in a first-person video game. If I was going to last six weeks in a tent with Roman, one of us would have to change, and it wasn't going to be him.

* * *

"We'll start here," said Roman the next morning, June 5, a few miles out of camp. He stopped walking and pulled a shell from the belly of the gun.

Five days of rafting and hiking had brought us—hips aching, egos smarting—to our first field site, beneath crumbling, windswept ridges that rose between five and six thousand feet above sea level. For most of its seven-hundred-mile length, the Brooks Range does not rise much higher than this, and—aside from a few heart-stopping subranges—resembles less the vertiginous shards of the Alaska Range than the jumbled bones of the Rockies. But these mountains are tall enough to shape plant life here. To their south, dense boreal forests stretch discontinuously for hundreds of miles, down to the coastal ranges of southern Alaska. To their north, open tundra rolls clear to the Arctic Ocean. Within these mountains rise the final ragged stands of the great forest: the northern tree line, the boundary beyond which trees do not grow.

But this line is by no means fixed. Twenty thousand years ago, back when what is now Seattle lay under a sheet of ice five times the height of the Space Needle, the entire Pacific Northwest was empty of trees. Ninety million years before that, there were temperate rainforests on Antarctica. Across Earth's history, forests have tended to migrate poleward when the planet warms and equatorward when it cools, pulling animals (and, over the last three hundred thousand years, humans) with them. But we'd come because the details of this migration—the rate at which it occurs, its causal mechanisms, and its net effect on global climate—remained murky.

One theory held that trees would eat up the tundra practically as soon as it got warm enough. Pollen fossils found in lakes suggest that forests repopulated the Far North rapidly after the ice age ended some thirteen thousand years ago, flooding north at rates up to twenty kilometers per decade. But pollen records are notoriously imprecise, and other scientists suspected that glacial refugia—gaps in the ice where pockets of hardy life clung to survival—might have endured here even in the depths of the ice age, inching outward once the climate warmed. If this were true, then the process of forest migration might be very, very slow. Journalists usually focused on the first, more dramatic hypothesis, but a 2009 meta-analysis showed that of 166 tree line sites monitored around the Far North, only a little more than half were moving at all. In short, the theories were flying, but no one had much more than a clue.

Now Roman held the shotgun shell up for us to see.

"Wherever this lands," he said, "will be the center of our first plot."

He tossed the round over his shoulder, and the rest of us dropped our packs at the edge of the trees. They were twelve- to twenty-foot-tall white spruce, *Picea glauca*, a hardy evergreen conifer with scaly gray bark and short, stiff needles—a worthy Christmas tree. Alaska's boreal forest also contains black spruce, paper birch, aspen, and balsam poplar, but white spruce are specialists of scraping together enough sustenance in summer to survive the desiccating cold of winter, and this far north they predominate.

Roman found the shell and led Russell, Duncan, Scott, and Felipe into the thin forest to document the older trees. Julia and I remained at the forest's edge. Our task was to survey the tundra beyond for young trees, called seedlings, to see if the forest might be on the move. A few knee-high seedlings rose before us, their boughs floppy and disproportionate, like the oversized ears of puppies. We documented these—their height, health, and surrounding vegetation—and then dropped to our knees, where we noticed smaller ones poking out from the undergrowth. Over the course of the next ten minutes, as we looked closer, the world closed in around us, and a lush, minute plantscape came into focus.

I had, to that point, imagined the Arctic tundra as a vague expanse, a blank spot on the map. What drew me were its negative qualities: the absence of human development, the freedom from distraction. The first several days of our journey had not much altered this view; I'd trudged behind Roman on autopilot, my senses disengaged, my mind conjuring all of the different foods upon which I intended to gorge myself in six weeks' time.

Yet what I found now was not an absence, but a pungent mesh of spongy mosses, neon lichens, and leafy greens. An entanglement of life that resembled, in its vivid color and complexity, a terrestrial coral reef. And as Julia and I looked closer still, a possible future came into view: Hundreds of tiny trees, a new generation no more than a few years old, bristled up through the soil.

Julia called Roman over. "Seedlings!" she said. "Tons of 'em." Roman looked skeptical, until I plucked one and passed it to him.

"Yup," he said, looking up with glowing eyes to survey the budding forest. "This is gold."

It was a thrilling moment, a first intimation that we might not be wasting our time after all. But I had no real sense for what it meant. Looking back now, I'm not sure even Roman did.

4

On June 8, after three days of fieldwork, we continued east to our second field site, blowing up the packrafts to float down rivers, rolling them up to hike over mountain passes. We cooked dinner over open fires and drank straight from creeks and rivers. McPhee had described rivers like this in *Coming into the Country*, rivers so clear that his eye passed straight through to their beds, "as if the water were not there." I hadn't fully believed this description before. Back in LA, the city's eponymous river flowed through a concrete embankment by my high school, and my friends and I used to joke, while peeing into the murky trickle during cross country practice, that we were disinfecting it. But here I was, dipping my cup and drinking snowmelt, which flowed like liquid glass over stones. Through the water darted bright silver fish: Arctic grayling, a species that can survive only in the coldest, purest streams and lakes.

As we walked, Roman discussed the implications of the boreal forest's advance, which, it soon became clear, were far more complex than I'd imagined. I had learned that trees were unequivocally good for humans; photosynthesis converts carbon dioxide into plant matter, removing planet-warming gases from the atmosphere. But this far north, Roman explained, trees don't grow large enough to remove much carbon. It's too cold, and often too dry. Meanwhile, their migration has all kinds of other effects.

One such effect is on the amount of sunlight the land reflects, a value scientists call albedo. A surface with low albedo, like asphalt, is dark and absorbs a great deal of solar energy, heating radically; high-albedo surfaces, like snow, are light and reflect most solar energy back to space. The albedo effect is so strong that if the entire planet were covered in white paint, its average temperature would drop to about −40°F. When ice and snow melt, the comparatively dark land or water beneath absorbs more heat, which then drives more loss of snow and ice—a self-reinforcing dynamic called the ice-albedo feedback. This feedback is one of the main reasons, aside from the greenhouse effect, that the poles are warming so much faster than the rest of the planet. It's also part of why the disappearance of summer sea ice in the Arctic Ocean gets so much play in the news: That change alone is expected to drive roughly 0.3°F of global warming over the next few decades.

The albedo effects of tree line advance are more complex than sea ice melt, though. Forest canopies are darker than tundra, and far darker than snow, which historically has covered the tundra through spring, when days are just as long here as in Miami. Trees absorb sunlight like black shirts, and then radiate heat down to the ground, melting snow. Most experts believe that the warming effect trees have by reducing albedo exceeds the cooling effect they have by turning carbon into wood and oxygen. But really, Roman explained, the net effect depends on a variety of factors, including how densely the forest grows, how snowmelt patterns change, and other dynamics we hadn't yet considered.

Sometimes Roman discussed these theories as we walked, but more often he directed our attention outward, toward clues in the land. Albedo was one of many relevant variables, and to grasp the bigger picture we'd have to think ecologically: considering not just isolated factors, but the entire environment and the way its constituent parts interacted. Roman was showing us the attentional habits we'd need to perceive these interactions. He was teaching us how to observe.

"See those mares' tails?" he said one day, pointing to wispy clouds reaching like tendrils for the sun. Formed by an incoming warm front pushing cooler air high into the atmosphere, the clouds heralded wind and rain in the coming days. Another afternoon, Roman asked, "What do you think those green patches down valley are?" I had no idea. "Dwarf birch," he said. "Bad walking." When we came across a recently downed spruce tree, Roman stopped and asked Felipe to saw the dead trunk into one-meter segments. Counting its rings, we found that the first meter of the tree had taken thirty years to grow; the second, eight; the third and final, just four.

"A meter in four years—that's pretty fast for the Arctic," mused Roman. "That makes sense. Growth is accelerating."

Other plants were growing faster, farther, higher, too. Especially shrubs—which, like trees, are woody plants, but shorter and squatter, with multiple stems. They too darken the tundra, and probably have many of the same effects on climate as trees. "I don't remember seeing any willow growing at this elevation," said Roman one afternoon, as we contoured up a hillside turning to willow brush. "This whole state is getting shrubbier." He had studied this phenomenon from his office, comparing past and present aerial imagery, but he'd also developed a five-tier brush classification system to apply on foot. "For Class Four, you need your body weight," he explained. "'Football brush,' I call it. You can break an arm in Class Four."

Roman pointed out wildlife, too: a pair of big-eyed boreal owls keeping vigil in the trees; loons floating on a mountain lake; a solitary wolf that slipped into a thicket before the rest of us saw it. He seemed to sense the presence of other creatures before they revealed themselves. This was what he loved about hunting, he'd told me the summer before: not the killing, which he did only for meat, but the quality of attention it demanded. "You're using all of your senses," he'd explained. "You're being quiet and your eyes are wide open. Your ears are open so that you can hear and find something. And you're looking at the tracks and really getting back to what it was to be a human, when we became humans."

For the most part, the young Alaskans took well to the new classroom, studying the land with patient intensity. Russell had taken a swiftwater rescue course from Luc Mehl, and on the river scouted shoals and hydraulics. Julia, whose earliest memories were from a five-gallon bucket her father turned into a seat and used to shoulder her into the wilderness, picked up rocks and told me their properties, plucked flowers and knew their Latin names.

My eyes, meanwhile, seemed magnetically attracted to the tops of my shoes. I was never the first to notice and was often the last to see. I took pictures and scribbled notes constantly; having been commissioned by *Smithsonian* magazine digital to write a story about the trip, I didn't want to miss anything. But while the theoretical stuff came relatively easily—Roman had downloaded dozens of papers onto two iPhones we'd brought for fieldwork, which I spent breaks and evenings reading—I found that during the long miles, my senses shut down. Memories of charged social interactions and unanswered text messages rose from the deep. I invented long and redemptive conversations with a high school crush with whom things had never—*why, exactly?* I now wondered—quite materialized.

By now I was beginning to find my way through the minefield of

Roman's equipment: I'd become gentler with zippers, refined my technique for getting in and out of the tent, and learned how to protect group gear from water damage. ("I know you've never been wet before, but show this stuff some respect," Roman had snarled on the third day, after I'd failed to adequately waterproof some batteries before rafting.) I'd learned, too, that I need not leave the tent to pee; our shelter had no floor, and one night, I heard a hiss and noticed that Roman had rolled onto his side: He was urinating directly onto the tundra next to his sleeping bag.

But if the first week had been a crash course in my practical incompetence, the second was a tour of my sensory oblivion. I wasn't just ignorant of the land, which came as no surprise; little wonder that where Roman saw signs of changing weather, I saw only pretty clouds; that where Roman recognized a natural growth record, I saw only a dead tree. No, the truly disconcerting part was that unless someone drew my attention to these things, they never even registered. It was as though I were moving through a separate reality. I proved myself capable of plodding along for hours at a time without noticing anything at all.

*   *   *

It is worth mentioning that I am not an unusually oblivious person. As a kid, I possessed that intense curiosity for the natural world that is common among children. In one of my earliest memories, my late grandfather is leading me around his Chicago suburb in search of cardinals and bluejays, downy woodpeckers and black-capped chickadees; he's slung a bite-sized pair of binoculars around my neck and is teaching me how to use them. For my fifth birthday, he bought me my first Audubon field guide, and together we spent weekend mornings birding along waterways and forest preserves, the sight of each new species stirring something deep in my chest. My family moved to Los Angeles when I was eight, and I was hesitant about the change until, on moving day, I spotted a hummingbird in our new backyard. *Maybe this place is okay after all,* I thought, recording the date next to the species in my new California field guide.

My obsessions at that age were short-lived, and I soon became more interested in tropical fish, which I encountered on a family vacation to Hawaii. My parents didn't share my zeal, but they nurtured it with more trips. I spent them snorkeling, diving, and documenting sightings with the fastidiousness of an aspiring marine biologist. The barrier between me and those colorful creatures sometimes seemed unbearable, like a rude encumbrance that I might, with enough application, eventually shake off. On one guided scuba dive, I spotted a small, bizarre-looking creature hiding under a coral head; when I pointed it out to the local guide, he shrugged. Recognizing the creature from my books, I reached for his underwater whiteboard and—hands shaking with excitement—scrawled:

curious wormfish

This pretty much stopped in high school. No doubt hormones played a role; who needs fish when Miley Cyrus is swinging naked into your bedroom on a wrecking ball? Then there were the AP classes, the sports, the college application rigmarole. But it was also around this time that I got my first smartphone. Our ancestors honed their senses to navigate the land in search of sustenance and shelter; by the time my friends and I entered the world, our attention had become a battleground for marketers seeking to exploit every cognitive foible, every evolutionary tic. By the time we started high school, this battle no longer stopped at the front door.

That was 2011—four years after Apple put the whole world in our pockets, a year and a half after Facebook created the Like button, and the same year Snapchat showed adolescents how to "live in the moment ;)." Almost overnight, the new digital culture flooded our lives, and only those things engineered to reach out and grab us—by our amygdalas, our dopamine receptors, our gonads—had a chance of getting through. Say what you will about the petabytes of vapid and often demoralizing content we consumed, but what about all we learned to ignore?

By halfway through high school, the quiet, in-between moments of life—the walks to and from the school bus, the free evenings after all my homework was done—had disappeared. I scrolled Facebook on my phone while watching YouTube clips on my laptop. Sometimes I'd catch myself typing into my browser's search bar the URL of the same buzzy website I was already on, like an amped-up lab rat chasing its own tail. It didn't feel good. Something wasn't right. When I encountered Henry David Thoreau's *Walden* in an eleventh grade English class, I thought I'd stumbled upon a holy text. "**Never has any passage reached me so intimately or corroborated so closely what I desire**," I wrote, trying my hand at nineteenth-century Transcendentalist prose in my iPhone's Notes app. A year later I went backpacking for the first time.

Those first trips were humbling, but it wasn't until entering the Brooks Range with Roman that the full extent of my oblivion came into focus. With each movement of his body, with each pause to listen, Roman gestured at a reality that I had mostly forgotten existed. At times my inattention verged on dangerous. I was lost in bitter thought, planning a searing rebuttal to some comment Roman had made, when the first grizzly of the trip wandered into our midst: Scott whispered, "*Bear*," and I tumbled back to Earth.

That bear, fortunately, trundled off at the sight of us. But the next one made more trouble.

5

think there's a good allometric relationship between the circumference of the anus and the size of the bear," said Roman, inspecting a pile of turd.

It was late afternoon on June 11. We'd just crested a mountain pass and were still several miles from the valley in which we hoped to camp, but Roman had paused because the turd before us was no ordinary specimen: it had been eight months in the making. During hibernation, fecal matter builds up into a so-called plug, which the bear passes upon emerging from its den. The passed plug before us was fresh and looked about the width of a fire hose.

"There's a big asshole out there," said Duncan, looking around.

"Sure hope we don't end up coming out of it," replied Roman. We'd been taking turns carrying the shotgun, and that afternoon Felipe was on duty; now Roman instructed him to move closer.

For the most part, bears are loath to attack humans unless surprised or aggravated. But by April or May, when the grizzlies of Arctic Alaska emerge from hibernation, they have not eaten for six to eight months. The caribou upon which they might otherwise prey have already migrated north to the Arctic coast, and the greens upon which they largely subsist will not emerge for another month or two. By early June they are ravenous. We'd passed valleys pocked with what looked like blast sites, where bears had moved a ton of earth for a bite of ground squirrel. They're so hungry, in fact, that Roman had broken with convention, instructing us not to make noise to alert them to our presence. "I'd rather sneak through," he'd told us.

I had no way of knowing whether this strategy—like our food arrangement—was unorthodox or unhinged. The bear safety literature says, emphatically and without exception, to make noise to avoid surprising a bear. Roman had taken a page from Teddy Roosevelt instead: "Speak softly and carry a big stick." Our voices might deter some bears, he said, but they would attract the aggressive ones. "The *moose* don't make noise!" he'd told me the first week, when I'd questioned his methods. "The only moose that make noise are the bulls, and the bulls get *hammered*!" So we moved silently now, following a well-worn caribou trail deeper into the canyon. Grizzly tracks the size of a baseball mitt gouged the wet soil.

"These tracks are fresh," said Roman after a mile or so, motioning for Felipe to hand him the gun. "We're walking into the wind, and so is he. I hate that."

I hated it too. It meant that the bear couldn't smell us, but it also rendered our bear spray—which blows wherever the wind does—worse than useless. And while the grizzlies of interior Alaska are far smaller than the behemoths of southern Alaska's coasts (who grow up to 1,500 pounds on the fat of salmon), adults here are still more than double the size of an NFL lineman, faster than Usain Bolt, and strong enough in the jaw to crush a bowling ball. You have only a few seconds, in many attacks, between visual and physical contact—not a lot of time

NORTH TO THE FUTURE

to stop a bear, no matter how good a shot you are. And that's assuming your gun works. The summer before, I'd met a biologist named Geoff Carroll who'd been charged by a polar bear on the northern coast. He raised his shotgun, but the gun jammed. Suddenly his twelve-gauge was no better than a Louisville Slugger. So he swung it for all he was worth, delivering a cracking blow to the charging animal's skull. The stock broke off the gun; the bear fell to the ground; and as it struggled back to its feet, another scientist arrived with a different gun, which he used as designed.

We had only one gun, and Roman didn't like it. His weapon of choice is a rifle; he likes to be able to shoot an aggressive bear from a distance. But if something happened to him, the rest of us would lack the marksmanship and composure it required. So he'd borrowed Luc Mehl's shotgun and loaded it with twelve-gauge slugs, devilish knuckles of solid lead that have less range but pack more punch. Roman didn't normally use shotguns, though, and he didn't like "the balance" of this one. And in the field, the action had begun catching whenever he rotated rounds.

"I have zero confidence in that gun," Scott said quietly to me as we walked.

"I feel like, under pressure, I'll be able to handle it," replied Roman, overhearing. "I only need one shot. The second one's gonna jam, but the first shot'll be good."

Evidence of the bear disappeared for a mile, and my mind began to wander again. But as we crested a slight hill in the canyon, Roman tensed; my vision tunneled. On the hillside below us was a big golden stud: *Ursus arctos horribilis*. A blond grizzly, digging for roots. Instinctively I switched on my recorder, as if documenting the encounter might help.[1]

---

[1] At the beginning of the trip, I received permission from the others to record at my discretion. All lengthy quotes in the book were transcribed from recordings.

At the sight of us, the bear lumbered a few yards away. But soon it stopped, turned, and dropped to its haunches, like a five-hundred-pound dog.

"That's a bad sign," said Roman gravely. "He doesn't know what we are, but he'd kinda like to find out."

We tried the usual tricks—making noise, huddling together, holding our arms up to appear large. The bear simply watched. Scott suggested firing a warning shot, and though Roman said it wouldn't do any good, he obliged, raising the gun and firing over the bear's head. The sound ricocheted through the valley: I flinched, the grizzly didn't.

"'What was that? Thunder?'" said Roman, imitating the bear in a slow, round voice. Seventy miles from the nearest road or village, he explained, the bear had probably never encountered a human, let alone learned to fear the deadly power of a firearm. "See, the sounds mean nothing," he added.

I was struck then by how calm Roman seemed. I was barely continent, and he was using the bear's reaction as a teaching moment. But while I detected no fear in his voice, there was a hint of sobriety, even tenderness. I would soon learn that he often gives voice to the thoughts of bears, caribou, wolves, and other animals. Even in potentially dangerous encounters—perhaps especially so—he tries to mirror the other's mind. Roman does not practice an organized religion, but he thinks he might have been a caribou in a past life, and would like to be a raven in a future one. "I hunt, but I'm not a hunter. I like using all of my senses; I like being quiet; I like cleaning the animal. I like feeding my family. But I don't like killing." He did not want to have to kill this bear.

After a few minutes, the bear stood, stretched, and began slinking across the canyon below us. Soon he disappeared into the trees.

"He's circling around to cut us off," Roman said, and began imitating the bear again: "'They seem to be afraid of me. Maybe...I can eat them.' Let's get out of here. Follow my footsteps *exactly*."

He led us down the canyon into a thin forest that limited visibility to distances the bear could close in an instant. As we moved in silence, I imagined a brown mass emerging over hillocks, bursting through brush. But what was cover for the bear was also cover for us—and since we were downwind, Roman's main concern was avoiding visual contact. He was trying to beat the bear at its own game: to sneak around it before it could sneak up on us.

This part of the canyon was filled with nondescript hillocks and look-alike gullies. After a few minutes of brisk walking, I no longer had a clear sense of where we were or where Roman thought the bear was. But Roman didn't slow to take stock of our surroundings. He moved like a wild animal: watching, listening, reacting, all at once. Of his skills as an outdoorsman, route-finding may be his greatest. While his adventure racing competitors were consulting compasses and drawing lines on maps, he kept moving. He read the landscape, letting the terrain and animals tell him where to go. "You get a lot of information walking on a game trail," he'd told me a few days before.

Some of this information he could convey—like the fact that bears like promontories, or that while caribou travel every which direction in the eastern Brooks Range, they primarily move north and south in the west. But much of it eluded description. We all looked for caribou trails, which make for quick and easy travel, but he invariably sensed them first. "I *know* where to go," he explained. "Sometimes people ask how, and I have to say, 'Well, hey, just trust me on this.'"

We were trusting him now. And when, after about twenty-five minutes of silent walking, we came to a clearing and he turned around to look, we did, too. What we saw deepened my trust.

On the ridge above, the bear had reappeared. It was sniffing the ground from which we had spotted it not half an hour before.

*   *   *

"It's not fun to walk away from a bear, is it?" said Roman softly as we continued. "They're worse trackers than most people think, but I want to sleep on the other side of the river."

We all nodded vigorously, I perhaps more so than the rest. Bears have more olfactory receptors than a bloodhound, and at the end of the day, the others would deposit all of their smelly things—peanut butter and bacon bits and sweating cheeses—in Roman's and my tent. Yet as we descended the canyon into a broad river valley, the land did not match our map. We expected to find a lake at the foot of the canyon, which we might have waded through to mask our scent and cover our tracks; instead we found an open basin with a thin creek gurgling through it.

The likely reason for the lake's disappearance, Roman quietly explained, was that the frozen ground beneath it had thawed sometime after 1983, when the US Geological Survey last updated its map of the region. The newly porous soil would have allowed the water to drain. This somewhat complicated our getaway, but it also shed light on another piece of the scientific puzzle.

The Arctic is underlain by soil that remains frozen year-round, called permafrost. Permafrost plays a pivotal but complex role in plant growth. For one thing, it indirectly provides water. Most of the Arctic receives less precipitation than Southern California; it appears lush only because permafrost traps what little rain and snowmelt falls at the surface, like concrete beneath a manmade pool. Warmer temperatures deepen the permafrost's "active layer"—the top section that thaws every summer and refreezes in the winter—allowing water to seep deeper into the earth, out of reach to vegetation. When enough permafrost thaws to drain bogs, ponds, and lakes, entire regions can dry out, becoming inhospitable to certain plants and animals.

In other ways, the thawing of permafrost *helps* plants. Tree roots can't penetrate permafrost; it's hard as rock. Only once the active layer deepens can trees colonize the tundra. And this unfreezing also yields the nutrients trees need to grow. Microorganisms wriggle to life, decomposing ancient plant and animal remains into nitrogen, phosphorous, and other necessary food. For these reasons, recently thawed soil can be fertile ground for new forests, if it remains wet enough.

But again, the story is more complicated. As we clambered over the far edge of the dried lake and into a thin spruce forest, we came to another feature that was not on our map: About one-third of an acre of ground had collapsed, resulting in a pit of fallen trees and murky water. Bone-dry spruce trunks hung over the rim like teeth; a fetid smell wafted up.

Permafrost holds the soil together, and when it thaws, the land can slump, slide, fall apart. Higher temperatures warm the soil gradually from the top down, but the results can manifest suddenly when pockets of ice in the ground melt. This process of rapid deformation, called thermokarst, creates structural problems for Arctic dwellers, and it poses a less visible threat to everyone else. Frozen ground underlies about 20 percent of Earth's land area and holds more carbon than all of the world's trees put together. It's all locked up in frozen plants and animal carcasses that have accumulated, layer by layer, for millennia. When permafrost thaws, the same biogeochemical processes that produce nutrients for trees also release this carbon as greenhouse gases.

One of the biggest questions facing climate scientists today is how much carbon this process—sometimes called the permafrost carbon feedback—will spew into the atmosphere over the coming decades. And unlike the gradual process of top-down thaw, thermokarst can abruptly expose meters of soil to air and sunlight, accelerating thawing and decomposition. The ice-rich areas that are most vulnerable to thermokarst also tend to hold the most carbon, and the water that often pools in them (left over from the melting of ground ice) triggers biogeochemical reactions that can turn much of this carbon into methane, a greenhouse gas far more potent than carbon dioxide. A well-regarded paper published shortly before we entered the field found that thermokarst might be far more prevalent than previously acknowledged, and that in the coming centuries it could roughly double the global warming effect of thawing permafrost—a possibility that climate models weren't taking into account.

Migrating forests matter, in part, for how they might affect permafrost. In summer, their canopies shade the ground, but their low albedo means they also absorb more sunlight, radiating heat downward. In winter, their trunks conduct cold into the soil, but they also trap wind-blown snow, which insulates permafrost from −40°F temperatures. (Snow, it turns out, insulates about as well as wood.) Limited studies suggest that the latter effect is the strongest, producing a net warming impact on permafrost. This is another reason that new Arctic forests *probably* accelerate warming. But as Roman explained, no one really knew.

Near midnight, still fleeing the bear, we emerged from the forest onto the bank of a river called the Wind. This body of water was very much intact: thick and sluggish with silt, it was more than three hundred yards across and looked like a sheet of flowing mercury. "We're not crossing that tonight," said Roman. "We'll try in the morning, when the water's dropped."

Reluctantly we made camp on the west bank—the bear's bank—

and waited for the sun to ease its work on the snowfields up above. As I settled into my sleeping bag, my thoughts returned to the bear, who was at most a few miles away. Recalling its slow, methodical gait and its half-open muzzle sniffing at the air, I knew that sleep would not come.

6

Six interminable hours later, when we broke camp to ford the river at its lowest ebb, there was still no sign of the bear. The river had dropped considerably; its broad silver surface was now pebbled by gravel bars, and though our legs grew numb in the icy flow, we emerged dry from the waist up. But a few miles later, I noticed a dark figure looming in a clearing ahead. The others, too, were silently eyeing the figure, no one willing to verbalize the apparition for fear of making it real.

As we drew closer, the figure turned an unnatural jet black, its angles unmistakably square: a cluster of fifty-five-gallon barrels. Our second cache of food and equipment. A collective guffaw poured from the group—too hard, too raucous, for the joke our eyes had played on us. I began then to recognize the psychological weight of the bear: how it prowled our minds, toying with perceptions and haunting our imaginations.

The barrels signified to us the opposite of the grizzly: safety and security. They would enable us to store food away from our tents and thus to sleep soundly. No less tantalizing were their contents—a fatty, sugary cornucopia to restore our dwindling rations. Peanut butter, instant ramen, canned chicken, canned cheese. Coffee packets, chocolate syrup, chocolate fudge frosting, plain chocolate. "I think chocolate is one of the food groups," said Roman one afternoon, as he finished his second Cadbury bar of the day.

Roman is, by his own account, "the opposite" of a kale-munching, granola-crunching nutrition nut. The summer before, after our moose hunt, I'd watched him polish off an entire pint of Häagen-Dazs in a grocery store parking lot. The key for long trips, he believes, is to eat energy-dense foods that taste good, even after weeks and weeks of the same. Calories are king. We'd all followed Roman's lead, stuffing our barrels with snacks that would have made us popular at a daycare. That afternoon we opened a can of chocolate frosting and scraped it clean within the hour. Our fires—in which we burned trash to minimize smell and weight—glowed and popped with the shriveling metallic packaging of Nabisco, Pillsbury, Pepperidge Farm.

We began referring to the day we reached a cache as "Christmas"—the key difference being, we agreed, that here we coveted fresh socks. (After two weeks without showers or a change of clothes, the odor inside our cramped tents was somewhere between "sweaty lineman" and "wet underwear mildewing in a trash bag.") We all packed small, light treats for ourselves, and this Christmas brought an extra gift: Roman had, earlier that year, written a memoir, and his publisher had sent an advance copy with the barrels for his review.

For three days, we spent the brightest hours studying the spruce forests near the barrels, and the evenings sitting around Roman, listening to him read his manuscript aloud. The fieldwork held few surprises; as at our first study site, the spruce here were growing riotously—plenty of recent growth, healthy needles, swarms of seedlings. But we were all intensely curious about Roman, who had undergone a striking transformation since entering the wilderness.

In the front country, he'd presented as a respectable (if unconventional) professor. One could imagine him roaming the aisles of a Safeway or cheering from the sidelines of a kids' soccer match. Now, dressed in tattered, decades-old layers, eyes glowing out of a fresh snowy beard, he seemed to be gradually shedding that veneer, and what remained mystified us.

His moods were mercurial, like the mountain weather, and it sometimes seemed as though the land and its inhabitants spoke through him. "It's gonna rain all day," he sighed one morning, after a few raindrops tapped our tent. I asked how he could tell; after all, he hadn't even looked outside. "It just has that sound," he replied, and was proved correct.

In some respects he was an open book. On the trail and around the campfire, he gave stream-of-consciousness voice to thoughts, volleying pet scientific theories and weighing navigation decisions, railing against oil companies and damning his aging hips. But the pathways of his thinking were so foreign to me, his sense of reality so rooted in phenomena I didn't see, that he remained largely inscrutable. Out of earshot, the rest of us talked about him constantly: *Why did he lead us to the ridge instead of the river? How could he have known that we'd find a game trail here? Why is he such an asshole? Why is he so generous?* And above all: *Why the hell did he bring us?*

This last question had nagged me ever since I'd received the invitation. The fieldwork required extra hands, but he could have picked seasoned adventurers. ("Man," said one scientist and experienced outdoorsman who had helped with our logistics, "if I could go on this trip, I would be right there.") Roman was nothing if not a shrewd judge of ineptitude, and he'd surely pegged me from the first for the greenhorn that I was. So why invite me to share a tent with him for six weeks? Why *any* of us? Since the moose hunt, I'd been cobbling together his past with an eye to this question, and now his book—first gripping, then lyrical, and finally heartbreaking—provided a missing piece.

*   *   *

Like so many Alaskans, Roman was not actually from here. Born in the suburbs of Washington, DC, he first visited Alaska at the age of nine, when his parents shipped him to the seventy-person coal mining town of Usibelli, in the interior, to spend the summer with his uncles. Only later would he realize that a troubled marriage had likely motivated his parents to send him away. When he returned, they had split up, and from then on Roman saw his emotionally distant father only on weekends.

But that summer was a revelation. His uncles spent their days in the mine, and it fell on Roman to cook, care for, and entertain himself. He took a correspondence course in taxidermy, acquired a .22 rifle, and roamed the surrounding creeks, forests, and mountains with a wolf-dog named Moose. He came for another summer when he was fourteen, and two years later, upon finishing high school, he returned for good. Having skipped a grade and graduated with straight A's, he was a strong candidate for any college, but he applied only to University of Alaska Fairbanks for its proximity to the Alaska Range: the biggest and some of the most rugged mountains in North America.

Soon after arriving, he fell in with an unruly band of alpinists who were making quick work of some of Alaska's deadliest unclimbed peaks. In the fall semesters, he'd earn perfect grades, distinguishing himself as a budding scientist and attracting the interest of professors; in the spring, he'd put school on the back burner to climb. The alpinist and writer David Roberts profiled Roman around then—the piece was titled "Roman Dial and the Alaska Crazies"—and described an idealistic rebel, a headstrong purist with a libertarian streak.

Roman wrote to the National Parks Service after it tried to shut down an early Wilderness Classic: "We are going to have a race. We have no permit. We won't even accept a permit if you give us one." In an essay called "Valdez Ice," Jon Krakauer—who was then a headstrong climber himself—recounted a young Roman leading him up a frozen waterfall so "brittle and insubstantial" that, for eighty vertical feet, he could not place a reliable ice screw to arrest himself should he fall. "Most people in his shoes would have been quite literally paralyzed with fear," wrote Krakauer, "which only would have hastened their demise. In Roman's case, however, the seriousness of the situation simply served to sharpen his concentration."

Several of his friends died in the mountains, and he might have, too, had not a few particularly close calls convinced him to swear off climbing at the age of twenty-six. He proposed to his now-wife, Peggy, and started a family. But while he completed a PhD in biology at Stanford and then took a professorship at APU, he by no means settled down. He traveled for research and still made harrowing personal trips, publishing accounts to supplement his teaching income. "Frozen Whiskey, Broken Skis, and the Taste of Death" recounted a 250-mile ski trip across the Brooks Range in winter. "Live to Ride, Ride to Die, Mountain Bikes from Hell!" told of a 150-mile bike traverse of the Wrangell–St. Elias Range. ("We carried no shelter, no stove, no luxuries," he wrote. "We'd come to ride, not camp.")

More often, he brought his family along on his adventures, determined to be more involved in the lives of his children than his father had been in his. He and Peggy had two children, a boy named Cody Roman and a girl named Jazz, and the four of them spent months in the wilds of Borneo, Australia, and Puerto Rico. Cody Roman especially took to the outdoors. "'His first name I took from Cody Pass, the wilder Alaska beyond Usibelli I had imagined as a kid,'" Roman now read aloud. "'Cody Roman, I reasoned, would be what lay beyond me.'"

When Cody Roman was six, Roman took him on a remote sixty-mile hike in the Aleutian Islands; it was on that trip that Cody Roman dropped his first name, adopting his father's: "Roman Two," as family and friends would come to call him. "R2." Later the two traveled the world together, rafting twenty-foot waterfalls in Mexico, searching for ice worms on Himalayan glaciers, and teaming up for adventure races. Like his father, Cody Roman studied biology in college and continued on to graduate school in the same discipline; by twenty-five he had published his first journal article. Two years later, in 2014, he embarked on a several-month adventure in Central America, the kind of grand odyssey Roman might once have undertaken.

It made Roman nervous. When Cody Roman emailed him plans for a ten-day solo trek through the Guatemalan rainforest, Roman drafted several discouraging replies but didn't send them. When Cody Roman told him of a rabies scare, Roman decided against suggesting he come home. "I've always resented people who warned me off my plans," he recalled telling himself. "How can I deny him his own adventure?"

After nearly seven months of bold exploration, Cody Roman set his sights on the Darién Gap, the narrow land bridge between southern Panama and northern Colombia known for its remoteness, deadly snakes, and *narcotraficantes*. To prepare, he decided to spend a few days hiking alone through Costa Rica's Corcovado National Park, and emailed Roman and Peggy just before setting out. "I'll be bounded by trail to the west and the coast everywhere else," he wrote them, "so it should be difficult to get lost forever." He was never seen alive again.

Here in the Brooks Range, Roman could be stormy and harsh. He seemed to find our inexperience exasperating, and after a lifetime of thirty-, fifty-, eighty-mile days, his hips were shot. As we traveled, he sang a rendition of the 1983 hit by Huey Lewis and the News, "I Want a New Drug": "I want a new hip / One that won't make me limp." But he softened in the tent most evenings, and one night he admitted to me that the thrill of adventure had obsessed him for much of his life. Now he sought to appreciate the things he had long taken for granted. He'd brought us twenty-one-year-olds along in part, he said, because he wanted to travel with people who were excited simply to learn from this place.

"I knew you'd be a little out of your element, but I could tell you were searching," he said, recalling his decision to invite me. "And when you get to be a certain age, you feel compelled to pass on what you know."

The book he read to us—since released under the title *The Adventurer's Son*, a national bestseller—recounted Cody Roman's childhood and the relationship the two had built through their shared passion for the natural world. The guilt he'd felt when his son had disappeared and the two endless years of uncertainty as he searched for him in the Costa Rican wilderness. The pain when he found the body, crushed by a fallen tree.

As he read the final chapter, Roman started making strange noises. It didn't sound like crying, exactly, but more like he couldn't get enough air into his lungs.

"I hope you guys never have to write a book like that," he finally added.

* * *

NORTH TO THE FUTURE

If reaching the barrels felt like Christmas morning, the day we left them was more like the beginning of Lent. The next leg of our journey would require covering a hundred off-trail miles in a week—16.7 miles per day, with one day off in the middle for fieldwork. Roman had planned each leg of the expedition to be longer than the last, and keeping apace would be partly a matter of getting fitter and partly a matter of cutting equipment. Preparing now for departure, we faced uncomfortable decisions: What to bring, and what to leave behind in the barrels? What did we need, and what did we merely want?

Our gear list had seemed to me marvelously streamlined when we'd left the road: We wore sleek trail-running shoes and shouldered gutted Dyneema packs with the aluminum stays removed. We carried space-age tents, five-pound boats, and three-ounce stoves. My pack—which held, in addition to camping and personal equipment, rafting gear (raft, paddle, life vest, dry suit), science gear (measuring tape, stakes, rope), and five days of food and stove fuel—had weighed about fifty-five pounds. Fairly minimal by most backpackers' standards.

But Roman was used to doing hundred-mile trips with little more than a few chocolate bars and a lighter. In his twenties, he considered tents a luxury item, siwashing like America's Natives, wandering the wilderness empty-handed like John Muir. Minimalism was both a practical and an aesthetic imperative: The less he carried, the farther he traveled and the more directly he engaged with the land. Roman's purism had tempered with age, and he now used two sleeping pads to cushion his hips at night; he'd even brought a comb, which imposed only a modicum of order on his unruly hair. But the day before we left the barrels, he had us each lay our equipment out in plain view of the others.

"The more we bring, the more time we spend schlepping, the less data we collect," he reminded us.

So we weighed the lightness of down jackets against the insurance of synthetic ones, which provide warmth even when wet. We balanced the utility of our communal pair of camp shoes—which let us air out our feet after long days in trail shoes—against the nutritive value of a bundle of chocolate bars. Things like athletic tape and bear spray canisters we realized we could share. I listed in my notebook the gear I left behind:

synthetic down jacket
long underwear (1 of 2 pairs)
extra batteries
ibuprofen
hand sanitizer
athletic tape
whistle
bear spray
watch

7

On Father's Day, June 16, the pilot picked up Scott and Felipe to return to Anchorage for other work, and the undergraduates—Russell, Julia, Duncan, and I—departed from the barrels with far less than we'd brought, following Roman northeast through a maze of unnamed mountains. The half-emptiness of our packs unnerved me; without my synthetic jacket, each dark cloud held menacing significance. Without bear spray, I felt naked. But Roman was right: As we dropped weight, my eyes seemed to magically unglue from my feet, and it became easier to look around. As our margin for error thinned, the land itself seemed to thrum at a higher frequency.

The thrumming, it must be said, came partly from the mosquitos. "We would have called it Utopia had not the mosquitoes nearly driven us wild," wrote the US Coast Guard officer John C. Cantwell, who explored the Brooks Range in 1884. "I have seen horses, fairly maddened by the torment, blindly charge through the forest, oblivious to the trees and branches encountered, until they wore themselves out," wrote the early-twentieth-century geologist Alfred Hulse Brooks, after whom this mountain range was named. "I have seen strong men... weep with vexation."

What mosquitos need most—pools of standing water—the Arctic's continuous permafrost provides more lavishly even than rainforests, and according to one entomologist's estimates, Alaska alone breeds some 17 trillion each summer. So bad are the mosquitos in parts of Alaska that the informal metric for measuring their severity is known as the "slap count," which is exactly what it sounds like. One hand, one slap, no smearing. Roman's personal record, set in the Brooks Range in 1990, was ninety-four.[1]

The other hateful troll of Arctic travel, no less ubiquitous or maddening, is the tussock. Composed of both dead matter and living, tussocks amass when a wiry cottongrass called *Eriophorum vaginatum* grows atop the undecomposed remnants of its predecessors, until it stands a foot or several above the ground, towering in unstable clumps that grow close together. From afar, an expanse of tussocks looks whimsical, like a sea of wispy pom-poms, but if one were to design an obstacle course to break ankles, it would be difficult to improve on tussock tundra. Nearly a century ago, the forester and conservationist Bob Marshall—the first modern scientist to study Brooks Range forests—described them at their worst:

> At least a hundred times in...three endless miles we would find ourselves sitting on the ground, a 65-pound

---

[1] The highest number I have ever heard, recorded by researchers near the Haul Road, is 278. Enterprising scientists have calculated that a naked human body, standing still on the tundra under such conditions, would die, in about twenty-two hours, of exsanguination. The caribou have no need for such thought experiments; they regularly lose up to half a liter of blood per day, and sometimes their newborn calves, to mosquitos.

pack anchoring us firmly in the mud, with an overhanging cliff of sedge formation nearly waist-high towering above us. We would grit our teeth, gather energy, and pull ourselves up the necessary three feet—only to do it all over again within the next twenty paces.

Yet neither the mosquitos nor the tussocks were as bad now as they could be. "Not even close," said Roman. That's not to say they weren't appalling; the sound of mosquitos whining just beyond our tent's bug net made me faintly claustrophobic, and many inevitably found their way in, driving me to sleep with my grimy face mashed deep into my now equally grimy sleeping bag. But on Roman's logarithmic scale—nine to twenty-seven bugs per slap is "moderate," twenty-seven to eighty-one is "bad," over eighty-one is "very bad"—Russell's leading slap of seven qualified as "low." And while the tussocks still spilled our feet from time to time and inspired much colorful language, they were, on the whole, relatively toothless. In fact, between them had begun to grow young blueberry, dryas, and willow shrubs: The tundra was turning to brushland.

The explanation for both of these reprieves, Roman suspected, was drought. Tussocks and mosquitos both depend upon standing water—and here the wells between tussocks were mercifully dry. One possible explanation for this dryness was that the permafrost was thawing, allowing water that had previously pooled to drain into the earth. Indeed, since crossing the vanished lake ahead of the grizzly, we'd passed the beds of several more ponds and lakes that had disappeared so recently that shrubs had only just begun growing in them. While doing fieldwork at our third field site on the eighteenth, I almost stumbled into a thermokarst sinkhole more than thirty feet deep.

But the tussocks were likely drying from the top down, too. The meteorological effects of rising greenhouse gas concentrations are wickedly complex, but the net effect, most scientists agree, is to reinforce existing weather patterns. ("Wet gets wetter," goes a common climatological saying; "dry gets drier.") And while the Arctic in general is dry, the southern flank of the Eastern Brooks is especially so, because it is bounded on all sides by mountains, which wring moisture out of incoming air masses. Not only does less precipitation fall, but warmer, drier air sucks more water directly from soils and plants, a process known as evapotranspiration. All of this can desiccate the land.

And drought doesn't just limit plant growth directly, by foiling photosynthesis; it also fuels fires, which burned two and a half times more acres in Alaska from 2000 to 2020 than from 1980 to 2000. Drought boosts plant disease and parasitic insects, too. The previous summer, en route to our moose hunt in Alaska's drying interior, Roman had noticed a brown coloration in the spruce trees we passed along the highway. "Let's see what it is," he said, pulling over. He got out of the car and grabbed a handful of needles; their tips were coated in a waxy substance that crumbled on contact, leaving orange powder on his hands.

"When I moved to Alaska in 1977, this kind of thing wasn't normal," he said. "But starting maybe two decades ago, there've been all these infestations of leaf miners, spruce bark beetles, fungal rusts. Something new every year."

I would have found it easy enough to spend the whole trip discussing these dynamics, but Roman had other aims. One afternoon, as we emerged from a field of drying tussocks onto a low ridge, he paused his discussion of drought and stopped walking. For the first time in several miles I looked up.

Before us stretched the confluence of two river valleys, each stretching north into a seemingly endless velveteen expanse. Empty of trees, roads, or any other familiar feature that would have provided a sense of scale, the view disoriented me: Whether it would take several hours to reach the horizon, or several days, I couldn't tell.

"It sure is beautiful," said Roman, looking down into the valley. "I feel this is wilderness."

If this wasn't, nothing was. The view was so big, so stunning, that I didn't know what to do with it. A part of me wanted to simply embrace its present glory, but I didn't know how. Another part wanted to capture and preserve the scene before it passed me by forever. Pulling out my point-and-shoot, I began snapping pictures.

* * *

In the second half of June, as we approached the two-hundred-mile mark of our journey, the loose daily schedule we'd initially followed went out the window. I'd left my watch—waterlogged from packrafting—in the last barrel, and the others stopped using theirs. The relentless sun, as we neared the solstice, joined the hours and days into a riverine expanse. We rose when Roman rose and camped when he camped, which generally meant walking when the wind died down and the sun was low, and sleeping when the wind picked up or the sun struck most brutally or it rained. "Like the bears," Roman told us. "You let the landscape and weather tell you when to go."

Losing the clock felt strange at first, but the land, Roman showed us, told its own time. "The growing season is so compact in the Arctic," he explained, "that one day here is like a week in the lower latitudes." We had left the road when the mountains were still purple with the stems of bare dwarf birch and the first willow buds had only just begun to sprout; now we watched the stark mountains turn green, then ignite with the purples and yellows and blues of dryas, lupine, heather; rhododendron, Labrador tea, anemone. Each day, a new wave of wildflowers shot out of the tundra, their stalks quivering in the wind.

We covered ground quickly now and took long breaks that felt like private islands of time. One afternoon we all swam in the deep pool of a mountain creek, rubbing weeks' worth of grime from our goose-bumped bodies. "Brrr!" blurted Roman, wading into the snowy water. Often we built fires and sat drinking hot water or coffee. Julia painted the plants and mountains with an ultra-compact watercolor set she had 3D printed for just this purpose, and Duncan, an amateur musician, strummed a fourteen-ounce carbon fiber ukulele that Roman had bought for him. Russell tended to study the map or sit quietly, looking around. Usually I scribbled notes or asked Roman questions, but occasionally I tried fishing with a cheap lure and a dozen feet of fishing line that I tied to my hiking pole. I had no idea what I was doing, and the others soon grew amused by my inept persistence. "Is it still called 'fishing' if you never catch any fish?" asked Russell one afternoon, like it was a philosophical question. *Casting*, we decided.

But the grayling sparkled like radioactive gems in the eddies of rivers—rivers so clear that the eye passed straight through to their rock beds. That water, those fish, stirred something I had not felt in years. They recalled warm childhood evenings on the lake in Wisconsin where my family spent several summers, and where I'd stand at the edge of our cottage's dock, casting for bluegills until the mosquitos chased me inside. I almost never caught anything, but whenever my uncle Mike came around, the fish began to jump. "Look!" he'd shout gleefully, pointing ahead; "Another one, look!"

Years later, I learned that those fish were actually rocks, thrown by Uncle Mike. That made more sense, because the lake was full of fertilizer runoff from the nearby farms and often coated in algal scum. But I remembered now what I had felt then—an excitement almost painfully exquisite. Each splash seemed a promise of infinite bounty, a testament to the world's grace.

8

As the weeks passed, a quiet settled over our group. The other undergraduates and I had spent the early days chatting nervously—discussing logistics, swapping bona fides. But now we knew our routines, and the pre-trip jitters had been burned for fuel, replaced by a dull ache that worked its way into our joints, bones, and minds. Chatter dissipated into the roar of the rivers and the stillness of the mountains. And in this new stillness it became harder to ignore certain facts: The indifference of the thin polar light. The strangeness of the rolling tundra, devoid even of trees. Out of camp once, while I was finishing digging a hole in which to do my business, a powerful gust of cold wind came, blowing my hat off, making an eerie liquid sound as it washed through the tussocks. I fought the urge to run back to camp.

This fear—of being small and vulnerable and almost entirely beside the point—is for me, and perhaps for most people, always somewhere in the background. It had been there when I was a twelve-year-old, watching a whale surface from a small boat in Hawaii. It had been there the previous summer on the glaciers of Denali. Sometimes it even crept in during long breaks from school, when the forward momentum of life slowed. But it almost never registered as dread or awe; I didn't open myself up enough for that. Instead it was a background anxiety I found ways to ignore, often with a notebook or camera.

The clearest instance came during the year after high school, on my first trip alone to the woods. I was eighteen at the time and had spent the previous nine months accumulating basic outdoor skills. That spring, high on the literary fumes of Jack London and Jon Krakauer, I hatched a plan to take a boat ferry to Alaska, park my car at the end of a road, and walk off into the bush—until my parents enlisted an older cousin, a skilled outdoorsman whom I deeply admired, to talk me out of it. (His winning argument: grizzly bears.) So instead I set my sights on Northern California's Lost Coast, the largest stretch of undeveloped American oceanfront outside of Alaska.

The idea was to spend two weeks hiking in the mountains, surfing along the coast, and shedding the trappings of modern life in order to commune directly with nature. But no sooner did I leave the road than I began filming flowers, spiders, trees—everything. I had brought my father's digital camera along with the intention of documenting my transformation, but I hadn't anticipated how strong the urge would be to put the world back behind a screen. Those mountains held miracles: enormous redwoods, herds of Roosevelt elk, elephant seals walloping each other with their blubbery bodies. But it was precisely when I encountered these things that the compulsion to film arose most forcefully. Had God appeared from the mountains, I would have filmed.

And of course, I filmed myself: pitching my tent, crossing streams, scrambling up hillsides. I'd set up the camera, step in front of it to march across the landscape, and then hustle back a few minutes later to retrieve the camera, like Les Stroud on *Survivorman*. I filmed myself filling my water bottle, sautéing vegetables, brushing my teeth.

NORTH TO THE FUTURE

"Give a small boy a hammer, and he will find that everything he encounters needs pounding," said the philosopher Abraham Kaplan, and there is probably a corollary for camera-phone-toting zoomers, an analogous urge to treat the world as so much raw material for social capital. But it goes deeper than that, I think. The selfie offered an almost irresistible solution to an alarming possibility, which the land raised with stubborn insistence: that perhaps I was not, after all, the unmoved mover, the axis about which the stars spun. The mountains, of course, did not vie for my attention; the flora and fauna seemed indifferent to my charms. For the first time in my life, I had no audience, no external reassurance that my existence mattered. So the camera filled that role. It was a simple, culturally sanctioned way to claim the place of central significance that I had rarely, in the course of daily life, had occasion to question.

The beauty and danger of the screen, according to the media theorist Thomas de Zengotita, is that it brings the world close, putting us somewhere near the center of things. While much of the internet's content is designed to trigger fear and rage, and though its algorithms tend to exploit our insecurities, the deep message of modern media—the underlying implication of all the dings and buzzes, all the outlets begging us to share our opinions, all the live events "brought to you," as broadcasters are so fond of saying, even on the toilet—is that we matter. Everything seems to be happening for our eyes, and in such a way that we can survey it from a place of invulnerability. We are free to merely spectate. When I momentarily lost this "God's eye view of the universe," as de Zengotita calls it, I clung to my devices all the tighter—diffusing charged moments, projecting myself everywhere. Systematically avoiding the kind of direct encounters that might actually have held the power to change me.

That's more or less what I was up to that first summer in the Brooks

Range. The *Smithsonian* article provided all the excuse I needed to cover the land in images and words, and as we walked, I scribbled in my notebook constantly, filling on average a dozen sweat-stained, trail-jostled pages each day. I kept a digital recorder in my pocket, and when it wasn't on, I wondered if it should be. I'd lurch ahead of the group or wait behind them to take photos, the digital *chcka* that mimicked a closing shutter reassuring me that the moment had been captured, stowed away for later review.

On the trail and in the tent, I peppered Roman with questions, some to do with the science, and some with what he saw in the land. "How do mycorrhizae affect tree line advance?" I'd ask. "How did you know there'd be a trail here?" Sometimes he tried earnestly to answer, but often he deflected, and occasionally he grew exasperated. "There are writers, and then there are authors," he said once, leaving me to decode his obviously deprecating meaning. And as the trip wore on, I began to sense the crudeness of my approach. **When confronted with a beautiful or strange sight, I noted one evening, I snap picture after picture and scribble note after note, capturing only details and abstractions. Roman just looks, but he seems to grasp the thing. It's like he possesses a set of faculties that I don't.**

Often I grew bitter in his presence and felt the need to assert myself, though I rarely had occasion. On the afternoon of June 19, as we entered the wilderness area of the Arctic National Wildlife Refuge—the northern portion of which oil companies had been targeting for decades and, under the Trump administration, seemed finally on the cusp of gaining access to—Roman veered into politics. My ears pricked up.

"I've been around the world looking for wilderness like the Brooks Range, and I haven't found it," he said. "Wouldn't it be silly to spoil this place just to make a few bucks? Wouldn't it be stupid to drill here just so oil companies can get even richer? And people who have gotten used to free paychecks can go on not working, buying a new snowmachine each year and letting it rust in their yard?"

"It's not that simple," I countered. "Basically zero Americans will ever get the chance to visit this place, but everyone wants cheaper gas."

"Would anyone die if gas were two cents more per gallon?" he replied testily. "Would you go on a road trip if it meant killing forty caribou?"

"No, I wouldn't," I said. "But I don't think that's the tradeoff. And there are environmental tradeoffs to pretty much everything we do. There are costs to having kids, costs to eating. Costs to this trip."

"That doesn't mean we should just throw up our hands!" he fumed. "You've got to draw a line in the sand somewhere."

A few days later, as we hiked up to a mountain pass, Roman made a barbed comment about the dubious value of cheap gas, and I reengaged him. We argued the merits of the permanent fund dividends, and the relevance of pristine wilderness to the average American; our obligations to other creatures, and the fact that high gas prices disproportionately harm the working class. I drew upon a sophomoric understanding of economics, environmental history, and moral philosophy to critique his position and articulate my own. Wielding words made me feel significant, expansive; and for the first time in days, I felt like a capable person. But words were not a game to Roman—not when they held the fate of the land. Twenty minutes into the conversation, he whirled on me, furious.

"It's so *annoying* to hear you say this stuff," he snapped. "Maybe you should fly out. Because if you can't appreciate this place, you don't deserve to be here." He kept walking.

I was angry and wanted to say something biting, but later that evening I felt like an asshole. The subject had not been alive for me in the way it was for Roman; it had remained speculative, like a college debate. Mostly I had wanted to hear myself talk.

* * *

Two days after that argument, a wolf began to follow us. It was skinny and dark and moved like a phantom through the open stands of white spruce, trotting and pausing to watch us, trotting and pausing.

We were cutting across the grandest valley I had ever seen, a complex of chert and limestone mountains that plunged thousands of feet into a large, clear river. The banks were sheeted in aufeis—ice that forms when ice-dammed rivers overflow in winter and then freeze, often lasting deep into the summer—and earlier, as we'd skated across these frozen slabs, a black bear had appeared across the river, spurring us onward. Now the wolf seemed a neat sighting, but I thought little of it; I was eager to climb out of the valley and into the highlands, where we could find wind relief from the mosquitos, build a fire, and rest.

But a few minutes later a second wolf appeared from the trees—bigger, closer. As the five of us turned to take stock, the second animal raised its head to the sky and howled—a sound more expressive, more baleful, than any I had ever imagined. A round, quavering moan:

aaARRRROOUUOOOUGHHRROOOOOOGGHHhhh

"They're packing up to come get us," said Roman thoughtfully. "They're collecting. They're hungry."

I didn't sense any fear in his voice; he seemed to be stating a simple biological fact. And as he finished speaking, another howl sounded in the distance—then another, then a chorus. Perhaps ten, fifteen wolves. It was hard to count, with their voices bouncing off the mountains, coalescing into a hair-raising song.

"Eerie," said Russell, grinning. I tried to smile back, but couldn't tell if I was succeeding; the howl seemed to have numbed my face. I knew that wolf attacks on humans were exceedingly rare; only two people have died this way in North America since the turn of the millennium, and both were alone. But I also recalled a chilling incident I had witnessed the summer before.

It had happened just outside of Utqiagvik, on a walk with the same biologist who'd fended off a charging polar bear with the butt of his shotgun. He owned a team of sled dogs and took six of them with us to rove the tundra off-leash. They were big, boisterous creatures, the kind of dogs that jump up to your chest and just about knock you over, leaving ropes of slaver on your hands. They were half Greenland husky and half Alaskan malamute, but "down underneath, they're wolves," the biologist told me.

Several locals joined us on the walk, including a couple and their two small daughters, the younger of whom was perhaps three or four years old. Shortly after we'd left the road, one of the dogs nipped at her. She yelped and began to cry. The mother picked up the girl, but the other dogs were already turning back toward the commotion—grouping up and loping around the mother and daughter in wide, slow circles. The biologist began jogging away from the scene and calling to the dogs to distract them, but the girl's desperate noises seemed to have flipped some primal switch in the dogs, and they continued to circle, closer and closer now, the mother calling out fearfully, the dogs yipping, a half-dozen animals all moving in unison, some clockwise and some counterclockwise, a blur of fur and teeth and raw instinct growing tighter and tighter, closer and closer, until the mother shrieked. The biologist yelled angrily, and finally the ancient spell broke. The dogs turned away from the mother and girl and began loping toward their master, tongues lolling happily. The young family hurried back to their car.

Later we learned that one of the dogs had ripped a gash in the mother's arm that had sent her to the hospital. It had all happened so quickly, the dogs moving so closely, that I hadn't seen the contact and couldn't tell which animal had done it. "Something happens when they start moving together," reflected the biologist back at his house, shaking his head, as he pondered whether he'd have to put the dogs down—the last dog team in Utqiagvik, a holdover from the time before snowmachines and four-wheelers. "The pack mentality takes over."

Now, as the wolves sounded through the valley, there seemed to be no space—only matter, pressing in close. It was as though my head flipped inside out: In rushed wind, sky, the smell of wet spruce needles. I was not merely watching or being watched. I was a body among bodies, a thing among things. I was *in* it. Roman lifted his head to the sky; his jaw went loose and his tongue lolled.

ARRRROOUUOOOUGHOOUGHhh

The sound was uncannily like the wolves, and I wondered what it meant. Was he communicating something to the pack? Was he just having fun? Seemingly in response, the wolfpack grew louder. Russell pulled back his head and copied Roman. Then Julia and Duncan joined, and so did I. I was surprised by how easily the sound came.

AAARROOOOOUUGGHHH

AARROOOUUGHRRROOUUUUGH

After a few minutes of howling back and forth with the wolves, I felt I knew what we were saying. Something like:

We hear you.

Us, too.

*   *   *

Later that night, after we'd escaped the wolves and retired to our sleeping bags, my thoughts remained down in the valley. I recalled the way the wolves had moved, and the terrifying beauty of their song. *It could swallow me whole, no problem*, I journaled, unable to define the *it*. I was relieved to be in the safety of the tent, but I also felt like something big had grazed me. Something I shouldn't allow myself to forget.

9

"Three," said Julia, looking up from the ground and massaging her craned neck. "Only three seedlings."

It was around four a.m. on June 25, and we were on hands and knees at our fifth field site of seven, in a valley called Red Sheep Creek, 220 miles of travel northeast of our starting point. At our first two study sites, we'd often counted fifteen or twenty healthy seedlings per square meter, and while most of these would surely die before reaching maturity, many might survive, establishing the new boundary of the boreal forest. Yet the farther east we'd traveled, the fewer we'd found—and without seedlings, forests can't migrate. For the time being, at least, eastern Alaska's tree line seemed to be staying put.

Whether this was a "good" or "bad" thing, I still couldn't say. Whether it was a surprise or not depended upon whom you asked, and when. Between 1804 and 1995, it would have seemed scarcely credible. In 1804 the Prussian scientist Alexander von Humboldt returned from a five-year voyage in the Americas, bearing discoveries that would lay the foundations of modern ecology. This was long before Charles

Darwin and evolutionary theory, in a time when most scientists were still, in the words of tree line ecologist Christian Körner, "busy cataloging Earth's inventory and separating the living world into labeled units." Humboldt took a revolutionary approach: He sought to understand nature as a dynamic, interconnected whole. "In this great chain of causes and effects," he wrote, "no single fact can be considered in isolation."

What Humboldt discovered, as he explored the Andes and the Amazon, is that there is a pattern to the distribution of plant species—a "geography of plants." By the end of the expedition, he'd collected some sixty thousand plant specimens and countless meteorological measurements, which he used to plot lines of equal average temperature, now called isotherms, onto maps of vegetation. These charts helped Humboldt establish one of the cornerstones of ecology: the theory that climate governs the distribution of plant populations across the globe.

This simple principle went unchallenged for nearly two centuries. In the Far North, Humboldt's successors found that tree lines closely followed the 10°C July isotherm, a line that wove in and out of the Arctic Circle.[1] As this temperature line began creeping north in the last century, scientists expected trees to follow. But in the 1990s, researchers studied several white spruce populations in the eastern Brooks Range and discovered that they were not growing faster than before. In fact, they seemed to be growing more slowly. Suddenly, modern science seemed to know roughly as much about forest migration as the traditional Inupiat of the treeless northern coast, who once believed that the driftwood washing up along their icy shores grew at the bottom of the sea.

"You would think: 'The trees are temperature limited; as soon as

---

[1] For a time, this was treated as a definition of Arctic, no less reliable than the latitudinal line 66° 34′N, which marks the southern limit of twenty-four-hour darkness and light at the solstices.

it gets warmer, man, they should grow!'" said ecologist Martin Wilmking, one of the early researchers of this phenomenon. "Well, they should. But we have also seen that that's simply not the case."

Our first order of business had been to see whether eastern Alaska's forests had begun marching north since the 1990s. Perhaps, in response to accelerated warming, trees had finally taken wing? Yet despite the rash of seedlings near the Haul Road, our field observations mostly refuted this hypothesis. What remained now was the thornier question, with which ecologists had been grappling for two decades: *Why?*

Two main theories had emerged in recent years, and the first was simple: drought. Perhaps eastern Alaska's forests were limited not primarily by air temperature, but rather water availability. If this was true, and if climate change was drying the region, then it followed that forests would stagnate.

And it seemed quite likely that climate change was drying the region. Not only did climatological theory support this, but scientists had used satellite imagery to discern the color of large swaths of the Far North, which provides a useful, if imperfect, proxy for net plant productivity. They'd found that the boreal forests of Alaska's interior—the vast center of the state, bounded by the Alaska Range to the south and the Brooks Range to the north—were not growing greener; in fact, they seemed to be turning more brown. Indeed, as we'd traveled east, the landscape had grown visibly parched. The tussocks were giving way to shrubs; the water-loving alders, relatively common farther west, had all but disappeared; and the colors of the landscape had faded. "This looks like it could be a savannah of Africa," said Roman on the solstice, June 21, as we prepared to walk up Water Creek, which was nearly dry.

The second, more complex theory held that the forests of the eastern Brooks Range weren't getting enough nutrients to push north—an issue that might be caused by permafrost. The soil here is colder than in most other areas along the 10°C July isotherm, because interior Alaska's winters—untempered by the ocean—are more severe. Even as the region warmed, the theory went, relatively little of this heat was being transferred to the microbes and fungi that produce necessary nutrients for trees; instead, much of it was expended in the thawing of permafrost.[2] If this theory was true, then the eastern forests would not access enough nutrients to begin moving north until much of the region had already thawed.

As we traveled, we explored both hypotheses—the first by gathering spruce needles, which Roman and his grant partner Paddy would later check for drought by analyzing variations in carbon isotopes. We tested the nutrients theory by installing soil temperature monitors and setting up controlled experiments at our study sites, fertilizing some trees and not others. (Later, Russell and a friend would return to our sites to see if the two groups had grown at different rates.) Yet all of this would take a couple of years to analyze. "All we can do right now is speculate," said Roman.

And of course, the possibility loomed that both theories were true—or neither. It was always a bit disconcerting to finish combing through a nine-square-meter swath of dense, matted tundra—composed of plants I was slowly learning the names and textures of—and then look up, across thousands of acres of unstudied land. To consider how many other factors might come into play a few weeks later, when waves of plants had bloomed and died, or a few *months*

---

[2] This would be due to latent heat, the energy released or absorbed when a substance changes states. When you leave an ice cube out on a hot day, for example, its temperature increases only to 32°F, at which point it hovers constant as the cube slowly changes from solid to liquid. Only once the substance has completed phase change does its temperature begin to increase again.

later, when the temperature had dropped by 100°F and a blanket of snow covered the ground and the wind howled like a banshee.

And this was only the surface. One afternoon, Roman dug a young seedling out of the ground and inspected its underside, stripping clumps of soil from its roots. "God," he said, staring into the intricate matrix. "Aboveground, it seems complicated, and then you look below ground and it's hopeless. One day they'll find out that the stuff happening aboveground hardly matters."

* * *

Meanwhile we continued traveling east, shedding layers along the way. We had packed for Arctic weather, but the temperature dipped below freezing just one night. Most mornings we were awakened by heat, which our tents trapped like greenhouses. Sputtering, we'd thrash out of our sleeping bags and stumble from the tents, damp with sweat.

Several months after the expedition, I described our experience to Rick Thoman, a veteran climatologist who had been watching Alaska's weather since 1987, when he began his career with the National Weather Service. Despite the utter lack of data east of the Haul Road, he made an educated guess, based on surrounding weather data and the topography of the region, that we had experienced the warmest period there in the state's short history. And the warmth would continue for weeks, culminating in a historic July, the warmest month in Alaska's recorded history (i.e., since 1925) by nearly a full degree Fahrenheit, and 5.5°F warmer than the twentieth-century average. Nearly two million acres of forest burned, releasing more carbon in a month than many countries emit in a year. Surface ocean temperatures off the coast of Western Alaska were 4 to 11°F warmer than average. For the fifth summer in a row, droves of seabirds—reliable indicators of the health of marine ecosystems—washed up dead along the shores of the Bering Sea.

"It's a pretty freaky time up here in Alaska," said Paddy, who had spent the previous spring conducting tree line research in the western Brooks Range. He makes this trip every year, armed with the knowledge that, due to disappearing sea ice, Western Alaska is now warming even faster than the rest of the state. Still, he was caught off guard when he arrived in Kotzebue that March to find the coastal town immobilized by wind-driven rain.

Across the state, March 2019 was 8.6°F warmer than the twentieth-century average, and in Kotzebue it was 21.9°F warmer. Paddy and his team survived the sixty-mile journey to their field site, but others in the region broke through thin ice and died. "People were falling through the ice left and right during a time of year when it's normally totally safe to travel," he recalled. "Everybody's historical knowledge was basically useless."

* * *

On the afternoon of June 25, as we walked between study plots, Roman suddenly seized up and dropped to the tundra. For more than two hundred miles, his arthritic hip had been grating, bone against bone, with every other step—and now his femur had slipped out of its socket. Cheeks sucked in, brow furrowed in concentration, he lay on the ground, staring past us into the valley below.

After a couple of minutes like this, he stood up and tried to keep walking, but a few steps later he sat back down. "This has never happened to me before," he said calmly. He instructed us to keep walking to the next field site, promising that he'd meet us there in a few minutes—and he did, his hip somehow back in place. But we all sensed that this moment had marked a turning point in the expedition.

The hip pain aside, Roman seemed, at times, troubled. "Where are all the birds?" he often wondered aloud. Each summer, a significant portion of the planet's avian population—some 5 billion birds—migrates to Alaska to nest in the all-day light. But Roman had noticed that the skies had quieted in recent decades, and they seemed particularly quiet here. In the tent at night, I was reading an electronic copy of *Two in the Far North*—the conservationist Margaret Murie's account of her legendary 1956 trip to the eastern Brooks Range with her biologist husband, Olaus Murie. I kept a list of the birds her party had seen, and on the morning of June 28, as we passed within a few miles of her camp, I read the list aloud to the others.

"Yeah, we haven't seen many of those," sighed Roman.[3]

Still, he continued to direct our attention outward, to the forested moraines of glaciers past, and to the fossils of corals formed hundreds of millions of years ago, before the Brooks Range rose from the bottom of the sea. He remained prickly and severe; though I'd stopped engaging him on conservation matters, his frustration with me continued to simmer. But still he taught us to identify louseworts, saxifrages, forget-me-nots; bug-trapping butterworts, and heliotropic dryas flowers. And while he continued to urge us to drop equipment, his encouragement became unnecessary. We'd learned that the less we carried, the faster we moved, the more we saw, and soon we were challenging ourselves to find out how little we needed. We used tape for bandages, dental floss to sew torn clothing, and cooking grease to oil the gun. "We'll just keep dropping stuff until we're naked and barefoot," said Roman one day, channeling a mode that the other undergraduates and I had begun calling "fun dad."

Despite all we had left behind, he encouraged Julia to keep her watercolor set, and we stopped occasionally for her to paint flowers, trees, and landscapes. Duncan held on to the fourteen-ounce carbon

---

[3] Three months later, a study published in *Science* found that since 1970, North America's total bird population had declined by roughly 29 percent.

fiber ukelele Roman had bought him and wrote a simple, wordless song inspired by all we'd seen. After dinner, we often sat around the fire for a while, singing.

And we encountered things, as we followed Roman east, that drew us out of ourselves: A mother bear shepherding her curious young cub, somehow rippling with fat even in lean country. Quarter-sized hailstones that drew blood from our knuckles as we held sleeping pads over our heads. A distraught ptarmigan hen clucking, feigning a broken wing, and zigzagging madly over hummocks and horsetail in a desperate attempt to lure us away from the vibrant center of her universe: a brood of downy chicks, blinking and quivering beneath a willow shrub.

In such moments, the visceral immediacy I had felt among the wolves returned. Attention, I had always supposed, was like a beam of light one endeavored, through an act of will, to shine upon worthy pursuits. Everything else was distraction. But the state of mind in which I now occasionally found myself was diffuse and responsive. I wasn't really choosing what to focus upon; instead I let the land reach out and ripple my senses.

Because what if I didn't *know* what was worthiest of attention? What if I'd never noticed it before?

10

On July 1, ten days before our scheduled pickup, Roman used our satellite phone to contact the pilot: He needed an early ride out. He'd pushed through another eighty miles of travel since his hip had seized up, but he'd begun taking anti-inflammatories during the day to function and Vicodin at night to sleep, and still the pain persisted. His trip was over, but ours wasn't. "You guys have passed the test," he told us. "Now you're ready for the final."

Before the trip, Roman had reasoned—based on the topography of the region and clues from satellite imagery—that the farthest north spruce population in Alaska must be in a valley about seventy miles northeast of our present camp. Somewhere along the frontlines of that forest would be the northernmost tree in America. "You just gotta go find it for me," he now told us. My chest, when he said this, cratered. But I had come to believe that Roman was so knowledgeable about the land, and so in command of his body, that even when chance events occurred—a bear trundling over a hillside, a storm thrashing our tent—it seemed that he was orchestrating them. A part of me now wondered if this exit, too, had been part of his plan.

On the eve of his departure, Julia wove a basket of green willow shoots, and we used it to cook pizzas over an open fire. Roman handed over the medicine kit, the satellite phone, and the shotgun to Russell, the designated leader of our final four-person push. "Don't run from the bears," he told us. "Don't shoot 'em either. You'll know if they want to eat you."

After the fire died, we all retired to our sleeping bags, and that night, just before waking up, I dreamed that I was returning from a voyage at sea. Roman seemed to be in charge of the ship, and as we stood on deck, pulling closer and closer to land, I was trying to thank him for something, a gift so exquisite that I wasn't sure I could bring it ashore. But for some reason—either my mouth wasn't producing sound, or Roman couldn't hear it—the words never seemed to reach him.

I was not sure that I would see him again. All summer long he had chafed against my inexperience, meanwhile inspiring in me an unflattering mixture of defiance and eagerness to please; a future trip together seemed unlikely. So over breakfast the next morning, I offhandedly mentioned my dream to Roman. He smiled. "If I go out east this year," he said, "although I'm not sure why I would—although I do like my editor in New York—if I go east, maybe you can help me avoid the tussocks."

Two hours or so later, the whine of an engine buzzed the air. Soon a plane came into sight, landing on a clear strip of river gravel, and a spry, elfin-looking man stepped out. He was the first new person we'd seen in five weeks, and I was struck by how clean he looked. We all hugged

Roman goodbye, and then he climbed into the plane and took off, disappearing over the mountains.[1]

Then we were alone.

* * *

The four of us traveled light and fast to the northernmost trees in America. We had by now left behind all but one tent, which we all squeezed into, and our packs were two-thirds empty. The chips and chocolate bars notwithstanding, we had lost weight, and our grimy clothing hung from our shoulders and hips. Our skin was scabby with mosquito bites. But our senses were coming alive to the land, and in Roman's absence we found our own route through the rivers and mountains.

Julia was particularly impressive in this regard. Her nose was so keen that she smelled animals and game trails before we saw them. "Caribou," she would say, and five minutes later the caribou would appear. Russell had turned twenty-two only a month before, but already he possessed rare composure, and when a large grizzly appeared ahead of us on our first day alone, I was glad he held the gun. Both taught me the names of more herbs and flowers.

"I still think I'm missing, like, 99 percent of what's going on out here," I told them one afternoon.

"I think that's true for everyone," Julia reassured me. "But I also think that just seeing a little bit, a little piece, is good."

And sometimes my mind quieted enough to notice the *chiggits* of

---

[1] Roman, it turned out, was not yet in the clear. On the flight back, the pilot recruited him to rescue two packrafters who had become stranded roughly forty miles away, on the Kongakut River. "You're the perfect person at the perfect place at the perfect time," the pilot told him. Despite his hip pain, Roman rafted Class III rapids alone, without helmet or life vest, to reach the stranded pair and shepherd them to safety.

the parka squirrels, or the pale sheen of upturned dryas leaves, which betrayed the comings and goings of larger animals. The cyclical rise and fall of the rivers, and the sun rolling hot and cold circles along the tops of the mountains. The winds blowing upvalley in the mornings and downvalley in the evenings.

Time, I was beginning to notice, passed differently here than it did back home. It was not just an abstract ticking that chopped life up into deadlines and appointments. Here time was movement itself, and we could see it in the changing colors of the tundra, and the smoothness of river stones. We could feel it in the daily exhaustion and recovery of our bodies, and the crunching of prehistoric marine fossils beneath our feet. It was the rattling hiss of the wind blowing through the willows, and the hillsides slowly sagging like fat as they thawed, and the decomposing caribou racks that littered the tundra. The white silence of the snowfields, and the deep pools that formed in the bends of rivers.

"There is no holding nature still and looking at it," wrote the early-twentieth-century polymath Alfred North Whitehead, who argued that reality was constituted not primarily by material objects, but by processes—processes defined by their relations to other processes. Each rock can tell a multimillion-year story of the mountains and ice and water that put it exactly where it is. Every plant hints at the local soil and climate. A single grizzly bear implies the existence of the million other nearby creatures required to sustain it.

I had never thought this way before—which is to say, ecologically. I had never thought to try. It would have seemed absurd to search for relationships between plastic chairs and disposable cups and mass-produced shirts, or to wonder at the origins of the images that flitted across my screens. Human desire was—particularly in the digital sphere—the organizing principle: We conjured things when we wanted them and disposed of them when we were done. But after five weeks of Roman's teaching and prodding, I was beginning to understand that there was a world out there that didn't care what I wanted. A realm of continual movement, beyond my powers of representation or even comprehension. A field of constant change, governed by dynamics far older than our species. And though it scared the hell out of me, I couldn't look away.

* * *

On July 8, after five days of travel and a morning spent walking up and down a valley on compass bearings, we converged upon the northernmost mature spruce tree in Alaska. Since Roman had first mentioned this objective, I'd been anticipating something dramatic—the missing piece to our scientific puzzle, maybe. But the tree, when we found it, was a scrubby little thing huddled in the depression of a hillside. If not for Roman's training, we would have walked right past it without a glance. Half a century old yet still coneless, barely six feet tall and hardly thicker than my wrist, it showed no sign of invading the tundra anytime soon. As to why the forests closer to the Haul Road were booming while these eastern ones remained static, we still had only ideas.

But as we took measurements of the tree, Julia noticed a dead chunk of wood near its base, where the crown had apparently died many decades ago, in the tree's youth. A branch had grown out sideways beneath the dead end, and then curved back up to the sky like an elbow joint, replacing the main trunk. The tree had almost died, but it had not—and now it was surrounded by several of its own clonal offspring, which its roots had sent up through the soil next to it, forming a grove of miniature genetic equals.

Julia's observation made us wonder at the conditions the tree must have endured here, all alone at the northernmost tip of its species' range. The gale-force winds and shuddering lightning storms. The dark, ungodly cold. Its haggard stature seemed a testament to something profound, though to what, I couldn't tell. "Life wasn't easy," observed Julia simply.

* * *

Three days later, our pilot picked us up near the Canadian border and delivered us to a small village on the southern flank of the Brooks Range. From there, we caught an eight-passenger plane back to Fairbanks. "You're welcome to get in," said the second pilot before we boarded, "but I recommend you all stay here." Farther south, fires were sweeping the state. "About half-mile visibility in Fairbanks," he warned.

Though I never seriously considered heeding the pilot's advice—we were out of food, after all, and had no cash to buy any from the village—I didn't want to go home, either. For the past two weeks, several wistful lines from Margaret Murie's *Two in the Far North* had been running through my mind: "We had a honeymoon," she wrote, "in an age when the world was sweet and untrammeled and safe." Her book mourned the passing of an old Alaska, before the oilmen and the slapdash frontier cities and the looming specter of climate change. An Alaska I would never know. I'd been hearing messages like these all my life, stories about how the natural world was slowly dying, and to some extent I'd internalized them. But while Murie's words moved me, her elegy, I'd decided, was not my own. All around stretched a world that, just six weeks earlier, I had scarcely known existed and was only just learning to see.

Soon after the plane took off, a wall of smoke appeared on the horizon. Plumes of ash billowed up from the smoldering landscape below. The windows paled to white, and then we could see nothing but each other and the inside of the plane hurtling through the sky.

PART II

SHIFTING FOUNDATIONS

> Let us settle ourselves, and work and wedge our feet downward through the mud and slush of opinion, and prejudice, and tradition, and delusion, and appearance, that alluvion which covers the globe… till we come to a hard bottom and rocks in place, which we can call reality…
> —Henry David Thoreau, *Walden*

> We have an *everywhereness* to us now…we are able to insubstantiate ourselves to the point that the solid stuff around us seems insubstantial.
> —Laurence Scott, *The Four-Dimensional Human*

1

The pilot was not wrong: The smoke, when Russell, Julia, Duncan, and I landed in Fairbanks, was too thick to spend more than a few minutes at a time outside. After seeing them off to the commercial airport to catch their flight back to Anchorage, I checked into a hostel in town, where I planned to wait until the fires died down enough to return to the wilderness. During those days, as I sat on my bunk bed watching ash swirl beyond the window, eating entire sleeves of Oreos and trying not to get sucked into my phone, I considered the multivariable complexity of forest migration. The careful, open-ended attention Roman had taught us to apply. And I began to wonder if these lessons might not speak to another scientific problem that I had bumped up against the previous winter, less than twenty miles from that very hostel. This was the problem of thawing ground—or, more precisely, *how to study* thawing ground.

From a distance it seems rather simple: The ground is frozen; as the atmosphere warms, the ground thaws; and in the process of this thawing, millennia's worth of preserved plants and animal carcasses decompose and release greenhouse gases into the atmosphere, accelerating warming. But in practice, studying this matter makes studying migrating forests seem almost straightforward, because forests, at least, are visible. Trying to study permafrost—which underlies about 10 percent of Earth's land surface area and takes decades or centuries to thaw—lays bare fundamental questions in climate science: questions about not just what we know, but what we *can* know. About our capacity to grasp problems so vast in their spatial and temporal reach that they elude direct observation.

To learn about permafrost, I'd arranged to spend part of that January, 2019, with an expert named Kenji Yoshikawa. But Kenji had his own ideas about what to teach me, and how. He believed in learning by doing. He'd left me alone for eleven days at his homestead, in near-total darkness and temperatures that plunged below −40°F.

Those days had been trying, and left me more confused about the science than I'd been going in. But seven months later, as I sat in my hostel waiting for the smoke to clear, I realized that Roman had given me a key to Kenji's perspective. The beauty of Kenji's work came into focus.

* * *

Kenji lived alone, off the public electrical grid, water supply, and road system in the coldest part of the state, the interior. Looking southwest from his cabin across the Tanana River Basin, you could see no other trace of humankind. Past his solar panels and artificial reservoir, the land rolled unbroken for dozens of miles—far enough that, even through binoculars, the distant stands of spruce and birch trees began to look like fuzz on a cotton shirt—before eventually heaving skyward into the granite-and-ice incisors of the Alaska Range. In the center of the horizon, 170 miles south and nearly four miles up, loomed the largest base-to-summit mountain in the world.[1] Variously translated from local Athabaskan languages as the Great One, the Mighty One, and the High One, Denali towers more than a half mile over the rest of the range, a twenty-thousand-foot mammoth without peer.

Kenji did not always live in such isolation. A world authority on permafrost, he was a professor at the University of Alaska Fairbanks (UAF), which lay only twenty straight-line miles from his homestead, hidden from view beneath the curve of his hill. After arriving on Alaska's shores in 1994, having sailed a small boat from his home in Japan up through the Arctic Ocean to study permafrost beneath the sea, he had embarked on a multidecade quest to chart the planet's frozen ground. This had involved skiing across the width of Greenland, snowmobiling from the west coast of Alaska to the Atlantic Ocean, driving across much of Siberia, and climbing many of the highest mountains in North and South America. As permafrost expert Larry Hinzman put it, Kenji had shed light on regions that were "previously considered unavailable. Or unavailable to…*normal* men."

But roughly a decade before I met him, he'd taken a step back from the world of teaching and publishing. Science's grasp on nature was becoming too theoretical for his taste; its practitioners were disappearing into institutions and computers, and losing touch with the world itself. So he'd turned his back on the academy, moved out to the homestead, and begun raising a team of reindeer to pull him, Santa Claus style, to research sites across the Arctic.

It was a lifestyle that allowed him to observe the land directly, free from many of the adulterating influences of modern technology. He could download weather reports to his phone—the prominence of his

---

[1] While Mount Everest is, at 29,032 feet, much taller than Denali (20,310 feet), Everest's base lies on the Tibetan plateau at about 13,800 feet, while Denali starts at about 2,300 feet.

hill afforded incongruously good broadband service—but he preferred to *see* the weather coming, and gauge inbound wind, rain, and snow for himself. On clear days, he watched clouds form and swirl across a landscape the area of Ireland.

I caught my first glimpse of Kenji's view in the summer of 2018, during my initial trip to Alaska. Having heard about him from other scientists, I sent a cold email while in Fairbanks, and a few days later he arrived at my hostel in a white pickup truck, wearing wraparound Ray-Bans and playing Led Zeppelin's "Dazed and Confused." In the passenger seat was a young man named Sama—a Mexican mountain guide and friend of Kenji's who was staying with him to learn about off-grid living.

After twenty-five minutes we turned off a two-lane highway onto a bumpy gravel road leading north into the woods. Shortly thereafter we turned again up a bumpier track, over which hung a red and black PRIVATE PROPERTY sign. The sign slid up the windshield and clattered over the top of the car. We passed another PRIVATE PROPERTY, a BEWARE OF DOG, and a custom sign that said:

PRIVATE ROAD

NOT HISTORIC TRAIL

NO PUBLIC ACCESS

About a kilometer later, Kenji parked the truck outside a small cabin and showed me into a canvas teepee he'd pitched behind it. As he began building an indoor fire over which to cook our dinner, I turned to business.

"So, are you seeing a lot of permafrost melting up here?"

It was an inelegant opening, but I was trying, at that moment, not to cry. There was an open smoke flap at the crown of the teepee, but the smoke was less rising than swirling—a fact which seemed not to bother Kenji, but which caused my throat to catch and my eyes to burn. It somehow felt important, in his presence, not to tear up.

"Permafrost doesn't 'melt,'" he replied. "It 'thaws.'" (*Melt*, I would later confirm, implies a phase change from solid to liquid, while *thaw* merely implies unfreezing.) And no, he said—in the two decades he'd been watching, Alaska's permafrost had warmed significantly, but relatively little had fully thawed. "There are some places, a little bit, but not much. Not that many."

The reply surprised me, as much for its tone as its content. Most of the scientists I'd met so far had been emphatic and unequivocal in discussing change in the land; they had a message for the public and were eager to share it. But Kenji seemed guarded, and when I said something to this effect, he looked up from the fire.

His was an imposing figure—impossibly broad in the shoulders, with a barrel chest, a thick neck, and a shaved dome of a head—but his face had so far been open and friendly. Now he wore a polite grimace, as though I'd brought a fetid smell into the teepee. He told me that journalists often asked him about climate change, but that he didn't give them the answers they were looking for. Much of the information produced by the broader climate science community he did not trust. He was not a modeler, he said—and good thing, because modelers were like pop singers: They got a lot of attention, but their work did not age well.

"I don't speculate," he said. "Just say what I got."

I had little time to follow up, because a car crunched into the drive: "The Germans," as Kenji referred to them, had arrived. In summer, Kenji's homestead is a lightly trafficked pilgrimage site for scientists, explorers, and others interested in the art of self-sufficiency. Already there was Sama, and now three young permafrost researchers, funded by an institute in Potsdam, had come to meet him. We all squeezed into the smoky teepee, and Kenji began cooking burgers over the fire.

As he watched the burgers, "the Germans"—only one of whom, I soon learned, was actually German—asked him about the fragments of his past they had heard: The time he'd skied from the edge of Antarctica to the South Pole. The six months he'd spent paddling *up* the Amazon River, alone, when he was twenty-three.

Kenji recounted some of the practical challenges involved in the latter trip: For drinking water, he scooped up buckets of the Amazon and left them on the sunny deck of his canoe for several days to disinfect. He brought no food—just a machete and a shrimp pot—and instead befriended indigenous locals, who taught him how to hunt and gather along the way. He particularly liked catching sloths, because they moved so slowly that he could keep them alive in the boat until he grew hungry. "I don't need the refrigerator or anything," he said, chuckling.

He told us about walking a thousand miles across the Sahara, from northern Algeria to the Niger border, to study the geology of the region. He had pushed a rickshaw through the sand with all of his samples and supplies in it, and packed only 1.5 liters of drinking water per day. Under the desert sun, his bread grew so dry that it left splinters in his parched mouth.

There was an unvarnished machismo to Kenji's manner. When one of the Germans asked if he knew one of their advisers, Kenji replied that he didn't, really. "We never drink together," he explained, as though this settled the matter. But he also exuded a gentle warmth. The thing he liked best about solitude and long expeditions, he said, was "freedom from thinking." During months alone, or while moving through a landscape for days and weeks on end, his mind eventually exhausted itself. "I don't need to think anymore, because I have already thought everything." Then, he said, his inner voice quiets. When he finally encounters another person, he feels something like love.

The burgers were cooked. Kenji handed them out on planks of two-by-four lumber, and the Germans—who, like me, were suffering in the smoke—received the makeshift plates with self-conscious grins. I was struck by the incongruity between the scientists: All studied the same natural phenomenon, and yet the well-groomed, fresh-faced Germans seemed to have arrived from a different planet than Kenji, with his beat-up sweatshirt, meaty hands, and quarter-sized frostbite burns on both cheeks. They reminded me superficially more of the climate scientists I knew from Princeton's Geophysical Fluid Dynamics Laboratory—scientists who worked mostly indoors, studying the natural world remotely.

After dinner, Kenji invited us all to drink with him and spend the night at his cabin. The Germans demurred; they had a meeting with their supervisor at nine a.m. I'd booked a flight to Anchorage the next day for Luc Mehl's birthday, so I decided to hitch a ride with them back to town. Kenji made coffee for the driver to help him stay alert—"Stronger always better," he said as he brewed it—and we left. (The driver later remarked that Kenji's coffee was indeed extraordinarily potent. "I'm totally wired," he said.)

But I left that night unsatisfied. I wanted a fuller understanding of thawing permafrost, a matter on which Kenji had an almost uniquely concrete grasp. I wanted to know whether he was right that a little thaw here and there wasn't so big a deal, and why he'd stepped away from the academy to become a reindeer herder.

And I see now that I also wanted a deeper kind of guidance. What struck me about Kenji was that he seemed to know exactly where he was and what mattered to him. Having grounded his work in the land, his life radiated a coherence that mine distinctly lacked. So that night I resolved to find a way to return to Kenji's for a longer stretch of time.

2

The first clear sign that the climate was changing, anywhere on the planet, came from the ground. It was 1986, and a bushy-browed geophysicist named Arthur "Art" Lachenbruch was conducting research on the North Slope of Alaska, near Utqiagvik. Lachenbruch was not your typical all-star geophysicist; a slow reader, he suspected that he suffered from an undiagnosed learning disability, and he had been told by his high school counselor not to bother with college. But over nearly four decades spent doing fieldwork in Arctic Alaska, he developed a personal knowledge of the land and a deep intuition for its workings, and—at the age of sixty—he came to a simple but profound realization: "The ground 'remembers,'" he wrote in *Science*.

He meant that frozen earth faithfully records past air temperatures, filtering out the noise of weather and preserving a clear climatic signal. Permafrost, by definition, holds no fluids that would otherwise pump heat into its depths; it's a solid, and thus warms only by conduction—a slow, steady process that takes years to flow meters deep into the earth.

NORTH TO THE FUTURE

By measuring temperature differences between successive layers of permafrost, Lachenbruch realized, scientists could render the most reliable information available about decades-long climate trends. "Careful temperature measurements made today," he wrote, "can provide information on the history of local surface temperature during the past few centuries."

By the late 1980s there was good reason to expect that the planet would warm, but few reliable, long-term weather stations existed near the poles, where changes were expected to register first. But now that Lachenbruch had discovered that the *ground* was a kind of natural weather station, he and a colleague gathered data from three boreholes in Northern Alaska—wells that had been drilled hundreds of feet into the permafrost to search for oil. They found that the soil there had warmed by at least 3.5°F over the previous century, and thus concluded that the climate in those places had warmed by the same amount. "That was the first irrefutable evidence that the climate was changing," recalled Larry Hinzman.

It was a brilliant discovery, but fundamental questions remained: Was the entire planet warming, or just part of it? How much of this change was driven by human carbon emissions, and how much by natural fluctuations in the Earth system?[1] How would this initial temperature shock reverberate through Earth's forests and oceans and ice sheets, and how much further warming might this drive in the decades and centuries to come?

In the years that followed, Kenji and his close friend and colleague

---

1 In fact, most of the warming Lachenbruch recorded was most likely caused not by human carbon emissions, but by a natural return to normal climatic conditions following the so-called "Little Ice Age"—a period of cooling that extended from roughly the sixteenth to the nineteenth centuries. Scientists have proposed several explanations for this cooling, including changes in ocean circulation and solar radiation, but the cause remains uncertain.

Vladimir Romanovsky continued to monitor Lachenbruch's study sites, finding that the soil there was warming by an additional 1°F to 1.5°F per decade, with the rate slightly increasing over time. And yet the above questions could not be answered by a couple of field scientists poking holes in the ground. Such questions, after all, were not about permafrost alone, but global systems. Not about particular places, but the entire planet, past and future.

Parallel to Lachenbruch and Kenji, a very different kind of science was emerging to address these questions. Variously called Earth system science or climate science, it drove advances in computing, modeling, and global systems thinking. It also quickly absorbed permafrost research as one of its subdisciplines. But the two approaches clashed. One was the domain of individuals and small teams; the other demanded sprawling, highly specialized institutions. One was immersed in the living, breathing world, and proceeded from direct observation and hard-won intuition; the other was more purely quantitative, and worked downward from theory and enormous stockpiles of aggregated data. At stake was not just a way of science, but a way of life: a vision of the world and our place within it. It was this conflict that drove Kenji off the grid—and in a way, it was also what drove me to him.

I knew little of this history when I'd first met Kenji, but that fall I studied abroad in London, spending my days working and scrolling in my cramped flat, and my weekends blowing off steam in noisy bars and clubs. Moving off the grid began to seem like a pretty good idea. In October I reached back out to Kenji. "I'm emailing to see whether you could use any help at the farm," I wrote, proposing to spend the first two or three weeks of January with him. "I'd love to learn the skills that Sama [the Mexican mountain guide] has been learning." What these skills were, I didn't actually know—and had I asked, I would

NORTH TO THE FUTURE

have learned that Sama occasionally looked after the farm when Kenji left town.

Kenji replied simply:

> Ok,
> It should work for both of us.
> See you in January!

3

The sky was dark as a cave when I landed in Fairbanks on January 1, 2019. Kenji was waiting at baggage claim, his bald pate gleaming in the brightly lit terminal. Outside, the air was surprisingly mild for winter in interior Alaska—which was, from all I had read and heard, not so much a season as a passage: a seven-month descent into metal-snapping, stone-shattering cold. But the temperature was now only a few degrees Fahrenheit below freezing, said Kenji, down from a few degrees above earlier that day.

As we drove out of the airport parking lot, I asked Kenji about a bizarre project he had mentioned the previous summer. It was an offhand attempt to make conversation, and I was not remotely prepared for his reply. The project involved a team of Japanese astronomers who were trying to transport the most powerful infrared telescope in the world to the top of Chile's eighteen-thousand-foot Cerro Chajnantor, where the thin, clear air would allow them to investigate dark energy and the origins of the galaxy. The astronomers had recruited Kenji's permafrost expertise to design a road on which to drive the massive telescope to the mountain's icy summit, and Kenji had agreed, but the project was not going well.

"Actually, I need to talk to you about that," he said, shifting in his seat. Recently another person working on the road had died, bringing the total death toll to seven. The astronomers needed more guidance, and Kenji had agreed. "I going back there next week," he added.

For the eleven days he would be gone, he explained, he needed me to take care of his reindeer herd, horse, dog, and property. I would have five days with him beforehand to learn everything I needed to know.

"Okay?"

Okay? I thought, my blood draining. *Does* okay *mean something else in Japanese?*

"Do you have medical insurance?" he continued. "In case something happens. Maybe I take picture of it."

We passed a car that had skidded off the road into a snowy ditch. About two hundred yards later we passed another. "Typical problem," said Kenji, explaining that the roads were icing over as the temperature began to drop. I was barely listening. Outside, fog shrouded the polar night, and the darkness beyond the road yawned without depth or focus. Leaving sun-drenched LA that afternoon, I had understood only dimly that my next eighteen days would pass in near-total darkness; now it dawned on me that my last glimpse of true daylight had passed during my layover several hours earlier, as I'd stared out absentmindedly over the Seattle-Tacoma tarmac.

Near the front of Kenji's homestead, about half a mile from the main cabin and teepee I had visited the summer before, was an older cabin that Kenji no longer used. As we rumbled up his old driveway, Kenji told me that it had a woodstove, and I was welcome to stay there. I swallowed uncomfortably.

"Actually I've never used a woodstove," I said. "And to be honest

I don't have a ton of experience building fires." By that I meant none. We didn't start fires in Southern California—we tried to put them out.

"Okay," he replied gently. "You stay with me in main cabin. I teach you before I go."

My thoughts lurched to the Jack London short story "To Build a Fire," in which a man travels alone through the Yukon wilderness, near Alaska's eastern border, in winter. After he breaks through a frozen creek, his feet begin to freeze, and he finds his fingers too numb to build a fire. London wrote two versions of the story, neither of which I now found particularly reassuring. In the first, published in 1902, the man eventually succeeds in building a fire but suffers frostbite that leaves him bedridden for a month. In the more famous version, published in 1908, the man has no such luck. "He realized that it was no longer a mere problem of freezing his fingers and toes, or of losing his hands and feet…The fear made him lose control of himself." Soon thereafter, the man falls into a slumber from which he never wakes.

A sickening anger rose from my stomach, directed as much at myself as at Kenji. In our first meeting, I had mentioned to him that I had climbed the north side of Denali, a grueling and remote route. Eager to gain his confidence, I had not mentioned how reliant I'd been upon the expedition's leaders, without whom I would have turned back or died. Now I felt sure that he had imputed to me skills I didn't actually possess.

We passed an old tractor, a stack of about fifty felled white spruce trees that Kenji pulls from in May to cut firewood, and a woodstove-equipped tent he had set up next to the driveway, in case the cabin burned down. Then we came to a gate, and Kenji stressed the importance of always closing it behind us: If a moose wandered in, it would be "almost impossible to get back out."

After another six hundred feet, we arrived at the main cabin, a twenty-by-twenty-foot box with a gable roof. Kenji had initially considered building his home underground to maximize heat efficiency and avoid property tax inspection by satellite, but decided it would be too expensive. Instead, he'd hired a contractor to build a house that was simple enough to complete during a brief window in spring when the days were long enough and before the ground turned soggy. The only parts of the stock design Kenji modified were the front windows: These he wanted to be very large to accommodate his view.

But the view now was darkness, and I had little interest in seeing it. What I wanted—what we both needed—was a drink. Kenji parked the car, and we trudged through snow to the cabin, the warm wind on our faces. Once inside, Kenji removed his flannel, revealing a T-shirt that said SIMPLIFY. From my bag I extracted a gift for Kenji, a half-gallon of Grey Goose. Kenji brought this to his liquor cabinet and returned with a fifth of Shackleton whiskey. Then, on the first night of the New Year, miles north of the coldest city in the US—which was, for the moment at least, rather warm—we drank, and drank, and talked deep into the night.

*   *   *

Much of what Kenji told me that night I didn't understand. His accent was thick, the material mostly unfamiliar, and I was too nervous in the moment to ask if I could record. The booze didn't help; even after my recent conditioning in the pubs of London, my tolerance was no match for Kenji's. So I scribbled down loopy fragments, and later—during the long evenings at the cabin and the months that followed—I pieced together the histories Kenji mentioned, and some he didn't.

He talked that night about one of his greatest inspirations, the father of western permafrost science, the great Ernest de Koven Leffingwell. Trained as a geologist, Leffingwell left his home of Illinois in 1906 to explore the northeast corner of Alaska, where the eastern Brooks Range falls gradually into the Arctic Ocean. He was the first modern scientist to study the region, and over the course of nine summers and six winters he traveled some 4,500 miles by foot and sled—dodging polar bears, befriending the Native Inupiat, and enduring temperatures so cold that he had to cover his metal instruments with surgeon's tape to keep them from ripping off his skin. Throughout his explorations, he meticulously documented a fantastical array of frozen landforms, and later he drew upon his observations to posit several theories that laid the foundation for the modern study of permafrost.

But those foundations were already on borrowed time. Four thousand miles away, a British mathematician named Lewis Fry Richardson was attempting, for the first time in history, to predict weather through pure mathematical calculation. Starting with a set of weather data collected across a region the size of Germany, Richardson first divided the area into a schematic of uniform, 200-by-200-kilometer boxes, and then interpolated for each box a single set of seven numbers: pressure, density, temperature, water vapor, and velocity in three dimensions. He plugged these values into a system of differential equations and began crunching the numbers.

It took him, by one estimate, roughly two years of calculation to "forecast" six hours of weather that had already long passed, and the result was quite likely the least accurate weather forecast in history: Nothing much happened that day, yet he predicted a surface pressure change of 145 millibars, equivalent to a weather event larger than Hurricane Katrina. Nonetheless, in the wake of this Promethean effort, Richardson published his dream of a "forecast factory," a hive of sixty-four thousand human "computers" who would calculate the planet's weather before it happened.

Viewed in isolation, Richardson's vision seems like the mad ravings of a megalomaniac, and yet it proceeded from a genuine problem: The atmosphere was—well, atmospheric. How do you study a phenomenon that is fundamentally global? "The meteorologist is impotent alone," observed the English polymath John Ruskin in 1839, decades before Richardson was born. In contrast to disciplines like physics and chemistry, he observed, the science of weather could not mature without a planetary view. The solution, according to Ruskin, was to construct "a vast machine": a global network of observers who would together form "one mighty Mind...one vast Eye...capable of solving the most deeply hidden problems of Nature." This was the call Richardson answered with his "forecast factory," and though human computers were too slow and fallible for the job, a more powerful tool was on the way.

In 1945, scientists contracted by the US military finished building ENIAC, the world's first ever electronic, programmable computer, and began using it for three purposes: modeling hydrogen bomb explosions, calculating artillery firing tables, and predicting the weather. The year before, the Allied invasion of Normandy had succeeded, in part, because Allied forecasters predicted a brief window of calm weather that the Germans didn't. Suddenly "meteorological mastery" became a matter of national security. Promising that weather prediction might one day lead to "weather control," the great computer scientist John von Neumann secured access to ENIAC for weather modeling. Over the next two decades, weather models and computing power grew in lockstep: The military gave meteorologists access to state-of-the-art supercomputers long before they became commercially available, and in return, meteorologists provided ever more accurate weather predictions.

And the science itself was changing, from an empirical discipline to a more theoretical one. Modelers were learning that the errors that inevitably arose in data collection could, when plugged into the nonlinear equations they used, multiply disastrously, distorting predictions beyond salvaging. (It was an errant wind reading, a meteorologist deduced, that had bedeviled Richardson's experiment.) To eliminate errors at scale, the data itself had to be checked against models and often "massaged" to conform to physical theory. This is not quite so sinister as it sounds; the task here was not to throw out information the modelers didn't like, but rather to comb through floods of incoming data in order to eliminate mistakes. And yet the cultural effect and philosophical implications were profound: Suddenly, meteorologists were answering to theoretical physicists and computer scientists, who called the shots about what data to collect and how. According to the historian of climatology Matthias Heymann, the new paradigm marginalized field scientists, ushered in a new era of "scientific elitism," and transformed meteorology from a process of "knowledge from below" to one of "knowledge from above."

And this approach would soon spread outward from meteorology to other environmental sciences. By the 1970s, the growing specter of climate change demanded predictions not just of the coming week but of the coming century—and because long-term climate is a function of oceans, glaciers, soils, and nearly every other natural phenomenon, climate science began pulling the other Earth sciences into its orbit, recombining the separate disciplines into enormous, office-bound modeling teams. In 1985, one geophysicist coined the term *Earth system science* to describe this new field, which treated the entire planet as a single, modelable system. Another scientist called this integration the "one model to fit all" strategy; a third hailed it the "second Copernican revolution." Richardson's "forecast factory" had grown beyond his wildest dreams.

This is, of course, an almost absurdly brief sketch of the new science; the history and workings of these models are too vast and technical to dispatch in several volumes, let alone several paragraphs. Yet this very incomprehensibility is itself a defining feature of the modern paradigm. "You end up with models where there isn't a single person who knows all the parts," one of my professors, a climate modeler, once told my class.

When John McPhee visited Alaska in 1976, he found a land of self-sufficient generalists—bush-dwelling men and women "of maximum practical application." These were holdouts of a bygone era, seeking independence from the rising tide of modern technology. But such people have become rare today, even in the hinterlands of the last frontier, and Kenji was one of only a few people I'd met who were still trying. He couldn't stand taking anything on faith. He insisted upon seeing the world with his own two eyes. Life inside "the vast machine" seemed to him like no life at all.

And as we sat there drinking, I realized that we were talking about *my* life, and I had only a faint notion of the alternative.

4

When I awoke the next morning around nine a.m., the sky was no brighter than it had been at a quarter past three, when we'd gone to bed. My head throbbed from the liquor, but Kenji was already up, silently tending to matters around the cabin with martial economy. If he was at all hungover, he didn't show it.

Kenji walks with his heavy arms bowed out at his sides like a prizefighter—which in fact he once was. In a 2012 biography called *Finding Mars*, the Fairbanks science writer Ned Rozell recounted how an eighteen-year-old Kenji was scouted by the renowned Japanese boxing trainer Shoei Uehara, who'd coached several world titlists. "I thought he could be a world champion," Uehara told Rozell. But while Kenji trained as Uehara's protégé for a little over a year, he was less interested in fighting than in disciplining his body and mind. "I think adventure was his dream," Uehara's wife told Rozell. "He didn't derail from it."[1]

---

[1] Many of the details from Kenji's childhood included in the next two paragraphs come from Rozell's excellent book.

NORTH TO THE FUTURE

Rozell's book described a man with an almost preternatural sense of direction. The son of an office secretary and a lifelong employee of Bridgestone Tires, Kenji grew up in the suburbs of Tokyo, the most populous metropolitan area in the world; Rozell visited his childhood home and found "no trace of wildness" beyond the manicured ginkgo trees lining the streets and the papillon dogs his parents doted upon. But when Kenji was five or six, his parents took him clam digging, and he slipped off to explore the shore alone. They eventually found him in the care of a bemused park attendant, who told them Kenji's surprising interpretation of events: "My parents got lost."

When he was nine, Kenji began sneaking out on bicycle, each time charting a new thirty-, forty-, fifty-mile route. In junior high he made a fifty-mile solo pilgrimage through the countryside on foot, sleeping in garages and train sheds. At the same age he learned how to build a fire, a skill he practiced every weekend, rain or shine. He studied astronomy, self-testing his knowledge by craning his head to a random point in the sky, allowing himself a one-second glimpse, and then naming the star he'd seen based on its color, the time, and surrounding constellations. These skills gave him confidence that he could always take care of himself, and always find his way.

"You have to believe yourself," he'd said the night before as we drank, pulling a sextant from the shelf and showing me how to use it. He lamented that people no longer saw the need to orient for themselves, since they carried phones that told them where they were. (I took a drink.) If you didn't trust your internal compass, then you were liable to become "lost."

I didn't need my English degree to understand that being lost, in Kenji's terms, meant more than geographic disorientation. He saw every device as a double-edged sword. To the same degree that technologies disburden us of practical concerns, the philosopher Albert

Borgmann argued, they also dissolve our concrete ties to the world. Even as they free us up to live a "life of the mind," they numb us to our physical surroundings. And in place of the focused, objective urgency one channels when building a fire against the cold, members of a technologically advanced society often suffer a more general anxiety, an ironic corollary of the fact that almost nothing we do on a daily basis is directly essential.

"Technology…I don't say bad, but changing everything last twenty years," said Kenji the night before, as he handed me the sextant. "I am last generation that knows how to use this. I feel obligated to share, or we lose everything."

Now, as I climbed down the ladder from the loft where we had slept, Kenji brought me a mug of coffee and began showing me around the single-room cabin, starting with the woodstove.

A good fire, he explained, begins with good wood, and good wood takes time. In May and June, after the snow has mostly melted and before the summer rainy season begins, Kenji gathers spruce and birch logs from his property, splits them into sixteen-inch quarters, and arranges them in neat, crosshatched stacks. He then leaves these to dry on a raised platform under a tarp for at least two and a half years. Behind his cabin were eight such stacks, and a few hundred yards down his hill lay another three. Each of these stacks—four feet wide, four feet tall, and eight feet long—is called a "cord," but Kenji warned that this term gives a false impression of equivalence. When he first moved to Fairbanks, he was told he would need to buy five cords per winter, but of his own thoroughly dried, tightly packed cords he finds he needs only two or two and a half.

Showing me one such cord now, his headlamp illuminating the crisply split wood, he pulled down an armful of billets, the frost from the back of the tarp dusting our faces. His armful was constituted, like the forests surrounding his cabin, roughly half of white spruce—a resinous, low-density wood that lights easily—and half of paper birch, a denser wood that burns long and hot.

Back inside, Kenji turned two valves on the stove to maximize airflow, and then opened the stove and placed two quarters of spruce next to each other to form a tight semicircle. Next he wedged a third piece on top. Tearing bark from one of the birch billets, he peeled a papery sheaf from its inside, lit the sheaf, and tucked the crackling flame between the logs. Within a matter of seconds, the spruce caught fire. Later we would add birch to the fire, reduce airflow, and close the bypass valve to force air through the catalytic combustor, all of which would increase the stove's efficiency. There was no smoke alarm, Kenji said, so I would have to monitor the fire myself. If I managed to burn down the cabin, I could retreat to the tent along his driveway until help arrived.

For power, Kenji explained next, he draws most of the year from three solar panels to light his cabin, refrigerate his perishables, charge his phone, and occasionally use his desktop computer. In winter, his solar power approaches nil, so he radically cuts down his electricity consumption, turning on lights sparingly and using his phone as little as possible. He converts his refrigerator back into an icebox and leaves his meat outside to freeze. By February, his solar panels will produce enough power that he can afford to be less frugal.

This was all fairly typical for rural Alaska, but Kenji's water system was more involved. For all its glacier-fed rivers and permafrost-enclosed ponds, Fairbanks receives less precipitation than Los Angeles. Kenji's homestead is one large catchment system. What little water hits his roof filters into six thirty-two-gallon trash cans connected along the bottom by PVC pipes. Any rain or snow that graces his land drains to an artificial reservoir he excavated down the hill from his cabin.

From the reservoir Kenji pumps water to his reindeer, but he also uses it for several less intuitive purposes. Every May, he stocks the reservoir with rainbow trout from nearby lakes. He is a skilled angler; about twenty rainbow trout can survive in a five-gallon bucket, and in half an hour of fishing his bucket is usually full. He fills another bucket with mud from the bottom of the lake, loads both buckets into the pickup truck, and—using an aeration machine to keep the fish alive—drives back to his reservoir.

He releases the fish and dumps in the mud, which is rich with shellfish and dragonfly larvae that would take decades for his pond to naturally develop in comparable quantities. "I don't want to wait that long," he says. "I just grab the existing ecosystem, bring back here." The upshot is that the fish and dragonflies eat pests that would otherwise plague his animals. Alaska's mosquitos are hellish, but they are by no means the worst insects around; warble fly *(Hypoderma tarandi)* larvae burrow into reindeer flesh, suck out nutrients, and burst forth nine months later, leaving oozing holes in their backs. (A single reindeer can host hundreds of these maggots.) The nose botfly *(Cephenemyia trompe)* lays larvae in the nostrils of reindeer, which then tunnel through its nasal cavities to the base of the tongue; there they grow so large that they can obstruct the animal's breathing.

But these appalling bugs are excellent food for fish, whose diet Kenji supplements with feed. In September, before the pond begins to freeze over, he catches the fish in two or three days. It is even easier this time because they have grown accustomed to his largesse; his fishing rod might as well be a gun, his pond a barrel. He smokes the trout behind his teepee and eats them through the winter.

"Good life," said Kenji, as he finished explaining the water system. "If the work not related to living, people feel more stress. Here, the work mainly related to living." But there was relatively little work to do now, he assured me. One benefit of coming in winter was that it would be too cold and dark to do much of anything, except look after the animals and cautiously consume the resources he had procured during the long, bright summer.

"You have to watch the fire, watch the electricity, watch the water," he summarized, turning to me and studying my face. I looked up from my notepad and nodded, trying to project confidence.

* * *

That afternoon, as the last of the pale January light died, Kenji put the finishing touches on a permafrost temperature sensor he planned to drill into the summit of Chile's Cerro Chajnantor the following week. He built his own instruments, he explained, "because don't really exist." Home Depot did not carry, say, a portable, battery-powered drill that could blast twenty-foot holes in frozen lava. So Kenji was, by necessity, as much an engineer as a scientist—and his fieldwork in the far-flung corners of the planet required him to be a mechanic, too.

Once, while studying subsea permafrost in the Arctic Ocean, his boat's engine malfunctioned in rough seas, and he had to disassemble it, identify and fix the problem, and put the engine back together, pausing to vomit over the lurching gunwale. Another time, part of his radio broke while he was skiing across Antarctica, so he transmitted a mission-critical radio message using an antenna he built from skis and other bits of metal. According to a former colleague, spending a day in the field with Kenji is "like watching twenty-four hours of *MacGyver*."

Kenji began his career in permafrost science when both he and the discipline were young. In 1986, the year Arthur Lachenbruch published his seminal work on permafrost's utility as a climate record, Kenji was a twenty-three-year-old at a crossroads in his life: He had already biked across the center of Australia, walked across the Sahara, and attempted to become the first human to walk to the magnetic North Pole—and as he looked out over the lifeless red dunes of the Atacama Desert, he decided that exploring for its own sake was selfish. He needed a higher purpose, and science would provide it.

Having already completed an engineering degree and published several scientific papers, he began a PhD, choosing to research the Arctic "because it was quiet," and permafrost because it was an undermanned discipline in need of his practical skillset. Following in the footsteps of Leffingwell, Kenji leveraged his penchant for adventure to produce maps of the planet's frozen ground where only piecemeal accounts previously existed. When he arrived in Alaska in 1994, almost all permafrost data in the state came from about twenty sites along the road system; Kenji proceeded to install hundreds of permafrost temperature sensors all over the region, and then did the same in Arctic Canada, Siberia, and several equatorial mountain ranges. As useful as Lachenbruch's boreholes were, they were prohibitively expensive to drill; Kenji's sensors provided an economical means of sampling a much vaster area.

"Around the world, he's given us this baseline of permafrost temperatures," said Larry Hinzman. "Now when you go back to those places and look at the permafrost temperatures again, you can see how much the climate has warmed in that location over that time period. It's a huge contribution."

But in the mid-2000s, Earth system science finally absorbed permafrost. Several studies found that hundreds of billions of tons of carbon might be locked up in the form of frozen plants and animals—and that as the Arctic thawed, much of this ancient matter might seep into the atmosphere. In 2005, *New Yorker* journalist Elizabeth Kolbert traveled to Fairbanks to shadow Kenji's good friend and colleague Vladimir Romanovsky as he checked the permafrost sites he and Kenji jointly monitored: "I think it's just a time bomb, just waiting for a little warmer conditions," Romanovsky told her. Then in 2007, a report by the Intergovernmental Panel on Climate Change (IPCC) identified thawing permafrost as a possible driver of abrupt climate change and one not adequately accounted for in models.

"Suddenly permafrost became very important," recalled Romanovsky, who remains one of the most widely cited permafrost experts in the world. "And of course, as soon as that happened, lots of people became permafrost scientists." He laughed. "Lots of good, smart people, but they don't have the basics. And that's always a problem."

The IPCC report triggered a flurry of research and news stories, not all of it properly contextualized. Articles often spoke of thawing ground as a novel phenomenon, when in fact, many regions had been slowly thawing since the last ice age. Stories blamed human carbon emissions for causing buildings around Northern Alaska to begin "leaning" and "sinking" into the thawing soils beneath them, but such thawing is more often caused by the structures themselves. Construction typically involves scraping away the vegetation and tundra that insulates the soil, and buildings, unless properly engineered for polar environments, tend to warm the land on which they sit.

Perhaps the most famous such building—a Fairbanks house that for decades has been gradually sinking, end-over-end, like a doomed ship—has become a poster child of climate change, but to date, all reporting on the house has overlooked the fact that the home suffered a significant flood in the 1980s, which likely triggered a rapid thawing process beneath. "I'm one hundred percent sure caused by the water leak," said Kenji, who had spoken with the owner about the home's history. "In general, almost all of the newspapers say something not correct. And the scientists let them! Just terrible."

As climate science pulls permafrost into its inner orbit, Kenji fears that his discipline is losing touch with its empirical foundations. The less directly researchers study the land, Kenji believes, the more arrogant and the less accountable they become. "Modelers who don't do fieldwork don't really understand," he told me. "Because they are only on the paper, reading, thinking—then making a solution. It's not true."

What they tend to overlook, according to Kenji, is the variety and complexity of permafrost, which is roughly as broad a category as "rocks" or "trees." Some is packed with carbon, and some holds very little. Some is full of chunks of ice that cause the land to fall apart when it thaws, and some is composed of materials that contract relatively little, so that the land slumps slowly and subtly. The previous summer, when I'd asked Kenji to explain how the many "drunken forests" around Fairbanks—stands of black spruce that lean and sag in thawed soil—had formed, I expected him to cite recent climate change, as several articles I'd read had done. Instead he grimaced. "You have to tell exactly where," he said. "All different."

To a certain type of person, half-truths and oversimplifications are even more offensive than outright lies. A few falsehoods can poison an entire discourse. Kenji had plainly developed an allergy to mainstream climate politics, but whether it was justified or not, I couldn't yet tell. He claimed that most newcomers to his discipline didn't really care about the Arctic. They flocked to the research questions that would attract the most media attention and grant money, rather than the ones most worthy of investigation. And most important, they didn't stay in Alaska long enough to truly understand the seasonal processes they studied, instead parachuting in during the warmest months of the year.

"They know only the summer," said Kenji. "*Of course* the idea about thawing permafrost comes easy." He looked at me and grinned. "If very cold wintertime comes, maybe thinking differently."

5

The next morning, January 3, Kenji had me start the fire myself. This turned out to be mercifully easy. Though I forgot to open the stove's bypass valve, the birch bark was so flammable and the wood so dry that soon a fire was blazing. *Maybe I'll be all right after all*, I thought, walking over to the kitchen to make coffee.

At the stove, I turned on one of the burners to boil water; nothing seemed to happen. Grimacing politely, Kenji rose from his armchair and explained that the burners didn't ignite automatically, but had to be lit by hand. I turned off the gas, and Kenji waved at the air above the stove. "When you go back to LA, you appreciate everything," he said, passing me a lighter. "Here, everything manual."

Once the stove was on and the water was heating, I grabbed a bag of coffee beans from the cupboard Kenji had pointed me to. Pouring some into his grinder, I hit power and the machine roared to life. Immediately Kenji leaped up from his chair and began waving his arms for me to turn it off. "Use pre-ground!" he exclaimed, and pointed to the cabin's battery monitor: its light had changed from green to yellow.

Grinding a few beans had depleted almost all of Kenji's power, and now we'd have to run the backup generator to recharge it. "Don't worry," he added, sensing my chagrin. "You learn everything before I go."

Once I'd finished making coffee—Kenji noticed the water start to boil, I didn't—Kenji motioned me over to his entryway, where he kept the car battery that powered his cabin. "Do you know about the inverter?" he asked.

No, I did not know about inverters. And I would learn little over the next few minutes, as Kenji began gesticulating and struggling to pronounce words I didn't know anyway—my practical knowledge of electricity being, if anything, inferior to my practical knowledge of fire. He was trying to explain how the inverter transformed the car battery's 12 volt DC current into the 120 volt AC current his appliances ran on, and why the generator would fry the inverter if both were on at the same time. All I managed to glean, however, was that if I failed to turn off the inverter before running the generator, I'd be hosed, and Kenji would be out $3,000. "Lose a lot of money," he said, frowning.

With the inverter off, Kenji told me to grab a headlamp and led me outside into the darkness. The temperature had dropped to 7°F overnight, and several inches of snow had fallen across the ice-caked ground. We crunched over the sparkling driveway by lamplight and descended into a subterranean cellar burrowed into the hillside, where Kenji keeps supplies he wants to insulate from the cold of winter and the heat of summer. He shined his headlamp on a red machine, about the size of the five-gallon fuel tank beside it. "You use before?" he asked.

"Uh, no."

Kenji began pointing to levers and buttons on the generator, then pressed two. "Choke on, switch on, ready to pull." He yanked on the pull cord with a beefy arm, and the generator juddered to life, its chattering racket reverberating off the frozen walls. He switched the machine off and let me try starting it myself.

Once I'd succeeded, we returned to the cabin, and Kenji continued explaining the charging process. "Run usually about two hours—*Oh, shit.*" Midsentence, the lights cut out; the cabin went tunnel-black. "Overload, I think," muttered Kenji in the darkness. What *overload* meant I wasn't sure, but it aptly described my mental state.

Kenji turned on his headlamp and began explaining the problem, and as I tried and failed to grasp his words, aware that my own incompetence had triggered this chain of events, my face grew hot. I wished I had actually learned the content of my high school physics class, instead of cramming it all into my head the night before exams, convinced that actually knowing things didn't matter so long as I had internet access. And it seemed strange: How was it that I used electricity every hour of every day without having any real idea how it worked?

Perhaps it was a sign of progress? "Civilization advances," wrote Alfred North Whitehead, "by extending the number of important operations which we can perform without thinking about them." Division of labor, economies of scale—the material abundance of twenty-first-century life would be impossible without them. But according to the philosopher-mechanic Matthew B. Crawford, something else was going on, too.

In his 2009 book *Shop Class as Soulcraft*, Crawford charted the disappearance of the mechanical arts from the American high school curriculum, as schools exchanged tool sets for computers and iPads. In eliminating woodworking, carpentry, and other such skills from education in favor of preparing the general population for so-called knowledge work, America seemed to be embracing "a vision of the future in which we somehow take leave of material reality and glide about in a pure information economy." Meanwhile the same general trend was taking place across much of the developed world—including, said Kenji, in Japan. "People under forty-five are dumb," he'd observed the previous afternoon, after explaining that in middle school he'd learned manual skills like how to disassemble an engine and put it back together. "They don't even know two-stroke, four-stroke." (I decided not to ask.)

This pedagogical shift was a symptom of a deeper cultural logic, according to Crawford. Not only was my generation learning to overlook physical things, but the workings of technologies were becoming increasingly obscure. "Lift the hood on some cars now (especially German ones), and the engine appears a bit like the shimmering, featureless obelisk that so enthralled the proto-humans in the opening scene of the movie *2001: A Space Odyssey*," he wrote. "Essentially, there is another hood under the hood." As computer interfaces "cover[ed] over" the mechanics of our everyday appliances, they were turning them into de facto black boxes, impenetrable to the layperson's inspection.

For the philosopher Albert Borgmann, the inconspicuous nature of technology was intrinsic to its function: The workings of devices were concealed in order to free us up to attend elsewhere. "If the machinery were forcefully present," wrote Borgmann in 1984, "it would eo ipso make claims on our faculties." Yet Crawford argued that the invisibility of machinery had, in many cases, become an end in itself. Engineers concealed the inner workings of devices in order to flatter consumers with an undue sense of control. The promise, Crawford wrote in his 2014 book *The World Beyond Your Head*, was that "the world will adjust itself to our needs automatically and the discomfiting awareness of objects as being independent of the self will never be allowed to arise in the first place."

Digital objects epitomize this promise, because they aren't even truly "objects." The noun *object*, like the verb *to object*, comes from the Latin *obicere*, which means to oppose or resist. Objects are *other*, confronting us with our limitations. The mountains, the snow, and the

NORTH TO THE FUTURE

cold are objective. Our bodies are also, at bottom, objects—objects that will one day decisively oppose our will to live. The objective world defines our subjectivity by resisting it.

But screens offer little resistance. Digitality reduces our hands to digits, tapping at what we want, exiting out of what we don't. Perhaps this is why we remember so little of our time online, and why an adolescence spent in front of screens often leaves us fragile and anxious. Standing there in the dark next to Kenji, I felt like I'd been shaken from a kind of dream logic. Like the syntax of my life was collapsing—which, in a way, it was.

By 10:40 a.m., the cabin's battery had finished charging, and we went back outside to turn off the generator. Fat snowflakes like ripped-up tissue paper were slowly falling. The temperature had dropped to –1°F.

* * *

By eleven a.m., the dim, spectral dawn had begun, and I could just make out the tops of the trees beyond the windows. Kenji led us outside and down to the reindeer, his life's focus for the past five years.

Kenji was the only person in all of Alaska trying to use reindeer as a form of transportation, and not for lack of would-be Santas. Eleven miles southwest of Kenji's homestead lies the two-thousand-person town of North Pole, named in 1953 to attract toy manufacturers. The manufacturers never came, but the theme stuck. The streets have names like Snowman Lane and Blitzen Drive, and for $9.95, the Santa Claus House—a tourist attraction that boasts a forty-two-foot-tall statue of Santa Claus—will sell you a one-square-inch parcel of bona fide North Pole property. But while the House owns several reindeer, they do not leave the Antler Academy of Flying & Reindeer Games. Products arrive by snail mail.

It turns out that there is good reason for this: Reindeer—which are simply domesticated caribou—are almost impossible to train. They don't even recognize their own names. To get one to pull him, Kenji scares it by running himself and then hops onto the sledge. He turns by using a stick to tickle the side of the reindeer's face opposite the direction in which he wants to go. The reindeer don't slow at turns, so he has to shift his weight to avoid flying off-trail. "Many times go straight," he admits; nearly every outing in his first two years resulted in a smashed sledge. When he wants to stop, he tugs on a rope attached to the animal and hopes it stops. Without instruction, and sometimes despite it, reindeer run to exhaustion—which means that if one takes off without him, it is not coming back. Once the harness wrapped around Kenji's ankle when this happened, and he was keelhauled for miles, losing most of the skin from his back.

Reindeer were first brought to Alaska in 1891 as a food source for Alaska Natives, after American merchants hunted whales and walruses—both dietary staples for the coastal-dwelling Inupiat and Yup'ik—almost to extinction. The missionary Sheldon Jackson argued that the most "practical" form of relief would be to ship reindeer from Siberia, and, incredibly, the federal government agreed. Over the following decade, as Jackson shipped more and more reindeer to Alaska, he became enamored of the idea of using the animals for transport, as the Sámi of Lapland had done for centuries. He predicted that reindeer would replace dogs in Alaska as the draft animal of choice, because while dogs had to be fed enormous quantities of meat, reindeer could survive by browsing the tundra for native lichens. Jackson proposed a vast network of reindeer mail routes, which he called "reindeer expresses," and by 1900, several of these routes operated.

But within a decade they'd all been discontinued. Jackson's estimates that a reindeer could "easily" travel a hundred miles per day dragging three hundred pounds of cargo quickly proved delusional, and early adopters found the beasts difficult to train and skittish around dogs. When the missionary Walter C. Shields—a man who proselytized reindeer travel no less fervently than the Bible—held a "reindeer fair" to celebrate the technique, the veteran herder Reverend Tollef Larson Brevig arrived by dogsled. "Why do you who are the oldest reindeer man on the coast come driving a dog team?" Shields asked. "Because I want to get there," Brevig replied.

While the use of reindeer as draft animals in Alaska largely died with Shields in 1918, Kenji believed that reindeer-powered travel had never been given a fair shake in Alaska. In Siberia—where the method originated, and where Kenji had visited roughly a hundred villages to study permafrost—Kenji had observed that sledges were inordinately heavy. "I want to make more lightweight, alpine-style," he said. "I want to see how fast the reindeer goes. Most traditional method in Arctic. I think someone has to do it."

Why someone had to do it I did not understand, but it told me something about Kenji. He believed in progress, but his version did not entirely coincide with technological innovation. It needn't be scalable or global, either; it was enough to work toward the modest resurgence of a traditional, sustainable method of Arctic travel, one that might tighten the bond between people and land, past and future. In this respect, his reindeer project was similar to his science: His primary aim was not to provide data to scientists in DC, but to put information in the hands of locals, who could interpret it as they saw fit. To this end, he trained villagers wherever he went in how to monitor his sensors; spoke at village schools; and even released a series of educational music videos, sung to tunes like Alicia Keys's "No One" and Gipsy Kings' "Volare." These activities had made him enormously popular in the Alaskan villages.

Whether or not I understood his reindeer project, my primary responsibility while Kenji was gone would be to look after his animals. Happily, early January was a relatively uneventful time in the life of a reindeer man. My duties were simple: I would feed the deer twice a day and keep an eye on them "in case something happen." I prayed to Saint Nick nothing would.

There were three holding pens, and the nearest two separated two bulls, each accompanied by cows. The third and largest pen held Kenji's castrated sled team along with several juveniles. Scooping feed into a five-gallon bucket, Kenji opened the third pen and led me inside.

From the far end, where the enclosure fingered into the surrounding forest, seven creatures appeared, streaming toward us through the dawn gloom. Reindeer are generally about half again the size of white-tailed deer, with a stockier build and disproportionately hefty antlers—the largest in the Cervidae family, aside from moose. They are handsome creatures, with coffee-and-cream-colored coats so thick that they do not melt the snow on which they sleep. A luxuriant tuft of white fur protrudes from their breasts, like a baroque lace jabot.

But I had never before spent time among large animals, and elegant as these ones were, their presence unnerved me. Their knees clicked loudly as they walked, an adaptation that communicates their movement and size to other reindeer, but which struck my ear as vaguely insectile. Their big wet eyes, as they drew closer, seemed dim and unseeing. Above all, it was the raw physicality of their group movement—a blur of mass and power, blind to all but the food in Kenji's hands—that bothered me. These creatures did not care to hurt us, I knew, yet this indifference was itself unsettling.

Kenji poured the feed into several bowls, and the animals swarmed around us, sticking their muzzles in to eat before Kenji could finish. The pellets clattered off their antlers and landed on their fur. As one of the larger males shuffled around me, I stepped back to avoid a tine in the eye. Seeing my dodge, Kenji nodded. "You just be careful," he said.

But if the reindeer gouged me, it would not be on purpose. The same could not be said of Kenji's sheepdog, Shiro, a white Maremmano of Cliffordian proportions. Shiro lived outside with the reindeer, serving as their first line of defense against wolves and other predators. To keep the dog's herding instincts intact, Kenji had not trained him at all. When I'd visited the previous summer, Shiro had growled and barked at me from his enclosure until I disappeared, at which point Kenji explained that he'd sent a previous caretaker to the hospital.

Now, as we fed the reindeer, I moved slowly and kept Shiro in view, trying and surely failing to project confidence. But for whatever reason, Shiro didn't seem to mind my presence. Sniffing my arm for a few interminable moments, he pulled back onto his hind legs, sprang into the air, and began circling me in airy, thumping bounds—pausing to eye me waggishly, his tongue lolling, his great body quivering in suspense. Then he'd launch off again, with a look in his eyes that could only be interpreted as pure delight.

"I wonder why he's so much friendlier towards me now," I mused to Kenji.

"Good mood, I think," Kenji replied.

6

"When animal come to the human, first contact is this property," said Kenji on the afternoon of the third day. We were in the cabin eating lunch—a hearty oxtail soup Kenji left perpetually simmering over the woodstove—and I was looking at a map of interior Alaska on my phone. Kenji's property is roughly in the middle of the state, and to the north of it, protected state and federal land stretches unbroken for 370 miles, to the edge of the continent. This is the largest wilderness in America, and as far as its nonhuman inhabitants are concerned, Kenji's homestead is part of it. "Good idea to check trail every day for fresh track," said Kenji. "Good to know if animal around."

In winter, when bears are in hibernation, the primary threat is wolves, which were largely responsible for the decline of Jackson's reindeer farms in the early twentieth century. Lynx can also grow large enough to poach young reindeer, and moose are wildcards. Once Kenji spotted a large bull nibbling at his plants and lobbed a snowball at it; the moose charged, and Kenji took cover behind a tree. For the next several minutes, as the moose tried to get at Kenji, and Kenji kept the tree trunk between himself and the half-ton creature, they got a good look at one another. "Eyes red," recalled Kenji. "Looks like drunk-next-morning kind of eye." Eventually the bull lost interest and went back to nibbling, but not before Kenji had learned a lesson. "Don't throw snowball in the moose," he advised.

We had spent most of the first two days around the cabin, and I had little sense for the rest of Kenji's 120-acre property. That morning we'd tried cross country skiing the trails Kenji had cleared for reindeer practice, but we hadn't crossed much country; roughly eight inches of snow had fallen over the previous forty-eight hours, and it was so bone-dry that it felt like sandpaper under Kenji's unwaxed skis. So after lunch, we lowered the airflow on the fire and suited up to groom the trails.

Usually Kenji did this by attaching a lightweight grooming mat to the back of his snowmachine (the Alaskan term for snowmobile). Atop the mat he would place bags of reindeer feed until it was heavy enough to compress the snow. But this method left much to be desired, because he had to remove the feed on uphill turns, and it often went flying off around sharp corners. Today we would attempt a new method.

On our way to the door, Kenji handed me two items. The first was a pair of rabbit fur mittens to replace the synthetic ones I'd brought. "Much warmah," he said. I didn't need persuading. I'd seen Kenji hold a steaming sweet potato in his hand with no sign of pain. He'd told me the day before that his fingers no longer bled when he cut them. His digits were so deadened from decades of frostbite that when he wears an oxygen meter for his high altitude work, he has to clip it to his earlobe, because if he wears it on his fingers it says he's already dead. I accepted the fur mittens gratefully.

The second item was more ambiguous. Reaching behind his couch, he pulled up the detached wooden leg of an old table.

Outside, the sun edged the horizon, glazing the treetops in coppers and golds and plunging the rest of the land into pools of deep shadow. The air was cold, and so clear that I felt as though I'd put on glasses. Steam rose from our mouths.

Kenji disappeared behind the cabin, and a moment later an engine roared; he emerged driving a snowmachine. Behind the vehicle dragged the thick grooming mat, about the length and width of an extra-large beach towel. But there were no bags of reindeer feed to weigh it down; back in the cabin, he'd proposed that I serve that purpose.

"Have you tried that before?" I'd asked.

"No," he'd said. "You first."

Now Kenji stepped off the snowmachine, tied a rope to the back of it, and paid the rope out a dozen paces. Then he took the table leg from my hands and girth-hitched it to the rope.

"Handle," he said. Stepping onto the grooming mat, he crouched and held the table leg like a wakeboarder. "You stand like this."

Trying not to think about what my mother would say, I took hold of the table leg and assumed an awkward squat. Kenji mounted the snowmachine and, looking back at me, eased us into a glide. The powder began to disappear under the mat with a smooth, plastic crunch; the sensation was so light and silky that I couldn't help but grin. We took two practice loops around the clearing, and then set off into the forest.

The trails were narrow, shadowy corridors, carved from a wood so dense that I could see no more than a dozen feet into it. Spire-like white spruce trees appeared to have grown downward from the sun. In the fresh snow were imprinted the comings and goings of creatures—meandering, circling, tearing through the dark.

Soon we had groomed over our ski tracks from the morning, and Kenji slowed and turned back to me. I raised one of my mittened hands in a thumbs-up.

NORTH TO THE FUTURE

"Okay, we going faster now!" he called back. "You have to—" and he mimed bobbing athletically, like a boxer dodging haymakers. Opening the throttle, he turned down a yet narrower trail, no more than four feet across.

The slope flattened, and the forest graded from tall white spruce to short, stubby black spruce. Trees listed drunkenly in the poorly drained soil, and a few leaned across the trail, forcing Kenji and me to duck as we zoomed through. Branches scratched my face and frost fell into my eyes. Occasionally we'd lurch over a fallen log, and my feet would leave the mat.

Kenji continued to accelerate, and as our speed increased, my field of vision narrowed. The world dissolved into a blur of wind and branches and flying snow, and soon I lost track of where in Kenji's maze we were. After some untold stretch of time, a small hillock, perhaps forty feet tall, appeared before us. I would later learn that this was a pingo: a bizarre feature formed when a pocket of subterranean water freezes and expands upward, swelling the land into an ice-filled bulge. We flew up the pingo, fell down its back, and began down a portion of trail Kenji deems too steep and narrow to ski. As we hurtled toward the bottom of it, I saw too late that the trail made a sharp dogleg right.

The mat made the turn; I didn't. In an instant I was flying toward the trees. Then I hit a thick snow bank, so soft and pillowy that it broke my fall painlessly. For a moment, the world was muffled and silent; then I emerged from the snow, jogged up the trail to where Kenji had come to a halt, and picked up the table leg. Off we went.

Some wind-blasted, adrenaline-soaked stretch of time later, we emerged into the clearing below the cabin, and Kenji brought us to a halt. We had groomed nearly all fifteen kilometers of his trails, and according to Kenji, I had proved myself more useful than a bag of barley pellets. "Person is the best," he concluded. For my own part, I still had little sense for Kenji's labyrinth. It seemed I would have to figure it out on skis.

As Kenji parked the snowmachine, I walked up to his deck and looked out over the woods. Forty thousand years ago, Kenji had told me, his hill was a sand dune, one of many that stretched across the Tanana Valley during the ice ages. (He could tell by the soil, an "aeolian silt" that had been ground up by glaciers and blown about by wind.) Back then, the Alaskan interior was too arid for anything, including ice sheets, to grow. It would have been a dusty, wind-ravaged waste.

Since I'd arrived, the view had been obscured by darkness and falling snow, but the sky had cleared over the course of the day, and now the sun—long gone below the southwestern horizon—cast a periwinkle glow over the Alaska Range, more than ninety miles south. Before me lay the largest unpeopled expanse I had ever seen, and every mountain, tree, and snow furrow stood in crystallized relief, as though chiseled from some enormous chunk of granite. A self-carving sculpture millions of years old and still in progress.

Whether from the overwhelming scale of the view or the cold, I wasn't sure, but my breath caught in my chest for a moment. I took a few more breaths, and then retreated into the warmth of the cabin.

7

It was sometime on my fourth full day at the homestead that the weather forecast for my time alone came into focus. Kenji was no great believer in forecasts; he preferred to watch the weather himself. "This is a big, big sky," he said. "Clouds coming or going, you can tell." But I didn't know how to read the clouds, and even a view the size of Kenji's can't tell you much about weather systems that are still half a continent away. And the forecast on Apple's Weather app was not encouraging: For at least the first half of my time alone, the weather would hover around −40°F.

It was only then that I realized I'd been mentally banking on the fact that winter here is not quite what it once was. Accounts of the Alaskan cold before climate change—accounts I had consumed, from the warmth of LA, with morbid fascination—now seemed almost too horrifying to contemplate. Accounts such as Barry Lopez's 1986 classic *Arctic Dreams*, which explored the polar night's reign over the local imagination.

> The oral literature of the Eskimo is full of nightmare
> images from the winter months, images of grotesque

death, of savage beasts, of mutilation and pain. In the feeble light between the drawn-in houses of a winter village, you can hear the breathing of something with ice for a heart.

Through colonization and statehood, the brutal winter has remained at the core of Alaska's self-conception, the central force against which the population defines itself. In one of his weekly columns, published in newspapers across the state, Ned Rozell put things concisely: "Cold is king here." Stories of the cold are, like stories of the bear, practically a rite of passage. Ask around and you will hear stories of cold so deep that the word shakes off its commonplace meanings, stories that tell what the thermometer alone cannot.

What does −30°F mean? Gelid fuel and solid vodka. A caught and frozen fish, when dropped on a hard surface in −30°F, can shatter like glass. When the temperature falls quickly enough, pockets of water trapped in stones can blow them apart, and sometimes the ground itself cracks with deep, resonant booms. Snow rarely falls—the air can hold almost no moisture at such temperatures—but when it does, it's so dry and light that it can form gravity-defying towers, piling eight inches high on telephone wires.

At −40°F, exposed flesh can die in five or ten minutes. A pot of boiling water tossed into the air will fall as a cloud of ice crystals. The deepest cold comes with perfect stillness, not whipping wind, because wind stops cool air from pooling. The atmospheric column often inverts such that temperature *increases* with altitude, sometimes by as much as 20°F in a hundred feet. The dense air layers trap sound like a tunnel; I have heard that in a deep, still cold snap, a quiet voice can carry for miles.

The coldest temperature ever recorded in the state was −80°F, in January 1971. Windchill is another matter. In March 1967, three men—Art Davidson, Dave Johnston, and Ray Genet—made the first ever winter ascent of Denali, often called the coldest mountain in the world, and as they began down from the summit, a fierce storm pinned them at 18,200 feet. For the next six days, as hurricane-force winds lashed the mountain, the men huddled in a snow cave they dug with the last of their strength—less than two feet high, and so narrow that their feet stacked on top of one another. When at last the storm abated, the men were wasted, half-frozen, but somehow still alive, and after staggering down the mountain they learned that the temperature, while they'd lain semi-conscious in the snow, had dropped to −50°F, and the winds had surpassed 150 miles per hour. Such numbers put the windchill they were exposed to as low as −148°F—an unprecedented figure that would become the title of Art Davidson's 1969 account of the expedition.

But less than two decades later, Lachenbruch published his findings on warming permafrost, and in the decades since, mountains of evidence have emerged from other scientific disciplines showing that Alaska is warming rapidly, and that winter, of all seasons, is changing fastest. Average winter temperatures across interior Alaska have increased by roughly 7°F. The number of days that reach −40°F—the magical number at which Fahrenheit meets Celsius, and around which mercury thermometers freeze—has dropped from an average of sixteen per Fairbanks winter to seven or eight.[1] Across the state, snowpack develops about a week later and melts two weeks earlier than it did in the late 1990s.

The texture of winter life is changing. Ice skating is a growing sport in Anchorage, because precipitation now often falls as rain, clearing ponds and rivers of snow. Winter rainfall is less fun for animals like

---

[1] This rough halving of −40°F days holds true even outside of Fairbanks and its potential urban heat island effect.

muskoxen, Dall sheep, and caribou, who must dig through snow to find food: If the rain doesn't thoroughly melt the snow before the temperature drops, it can form an impenetrable sheet of ice. Sea ice forms in the Bering Sea two to three weeks later on average than it did in the 1980s, and melts two and a half weeks earlier in spring. The winter I visited Kenji, it didn't form at all. That March, competitors in the iconic Iditarod sled dog race would—for the first time in race history—cross the finish line in the coastal village of Nome with an open ocean behind them.

"Lately, weather, our favorite nemesis, has broken the rules," wrote Alaska's foremost novelist, Seth Kantner, who grew up hunting and fishing from a sod igloo in Northwest Alaska. "Our confidence in the most trustworthy feature of the Arctic, winter, has been wounded." Such sentiments have become legion in Alaska. Nine months before I visited Kenji, the Grammy Award–winning composer and longtime Fairbanks resident John Luther Adams had published an essay titled "The End of Winter," in which he observed that the fearsome Alaskan cold had begun to seem oddly fragile. "For me, as for so many others, Alaska had been…the last great untouched place in North America," wrote Adams. "It's become one of the most threatened parts of this increasingly threatened Earth…as the ice disappears and the waters rise, what will take the Arctic's place in the human imagination?"

Such accounts had struck me forcefully in LA. I loved Jack London's stories of the Yukon winter and had often tried to imagine myself inside of them. *How would I fare? Could I hack it?* It somehow felt important that the deepest cold continue to exist somewhere on the planet. Simply knowing it was out there fired my imagination, even if I never experienced it myself.

I did want to experience it, though—or so I had thought. But as Kenji prepared to leave and the temperature steadily dropped, I found that I was not wanting for imagination. My mind flickered with appalling possibilities, and the end of winter was not one of them.

*   *   *

That evening, two days before Kenji's departure, we decided to take advantage of clear weather to stargaze. Down on the pingo, he had set up a clear plastic dome years ago for this very purpose. He typically warmed the dome with a portable space heater, but when we snow-machined down to it and crawled inside, Kenji found that the heater wouldn't start. The temperature up at the cabin had been −16°F; down on the pingo it must have been five or ten degrees colder.

As was the case most days, we'd been drinking beer through the late afternoon and evening, and now we drank vodka to keep the warm tingle going. Kenji also fired up two gas-powered stoves to heat the dome. Still, we grew cold lying there, and after a while we found that the bottle of vodka sitting next to us had frozen solid. Kenji tried fixing the heater, and I went outside to take long-exposure photographs of the stars.

I took off my mittens and held still to work the camera. *Snap. Snap, snap.* The battery on the camera drained quickly, and after fifteen minutes or so I found that my fingers fumbled at the buttons; I had to press down with my whole wrist to close the shutter. Then I realized that I couldn't feel my toes. I tried wriggling them but could find nothing to wriggle.

Working to keep my voice level, I called out to Kenji to suggest that we head back to the cabin. He agreed, and while he packed his things, I jogged around the pingo and stomped my feet. My toes began to hurt, which I knew was a good sign. *Everything's fine*, I told myself. *We're less than half a mile from the cabin. It's not* that *cold.* But these rational sentiments did not feel like me anymore; they were flimsy, floating things, and beneath them swirled an animal fear. I could sense the cold creeping into my body, reaching for a grip on my brain. Suddenly I saw how easy it would be to lose control: A stalled engine, an empty lighter, and the night would close in.

Kenji fired up the snowmachine, and I clambered on behind him. As we sped through the forest, I felt like a child clutching a parent.

"Next week," Kenji told me later that evening, "you better be careful with frostbite."

8

The following evening was Kenji's last at the homestead, and we drank heavily. Kenji believes in the Japanese concept of *nomunication*, a portmanteau of the Japanese *nomu* for drink and *communication*. "If you don't drink," he told me, "you never success in business." We began with a couple of beers each, and then moved to Bulleit Frontier, Macallan, Martell, Jack Daniel's 128 proof, and eventually moonshine. By eight p.m. the room was pulsing, and I turned on my recorder. It was time for business.

Since arriving, I'd been looking for chances to discuss thawing permafrost in more detail with Kenji. He found climate talk distasteful, but I was not going to let him leave me in the dark for eleven days without squirming a bit. As we drank, I pressed him on the subject of an article published in *National Geographic* a few weeks after my summer visit.

The article, written by the respected science writer Craig Welch, focused on a potentially grave inflection point in global climate change, which seemed to be approaching sooner than expected. When scientists and policymakers talk about permafrost thawing, they are usually referring to the so-called active layer thawing more deeply in the summer, absorbing more and more of the soil beneath into seasonally thawed mush. But for several years now, this top layer had remained active late into the winter in parts of the Arctic and subarctic, and in the winter of 2017–2018, several study sites near Fairbanks never refroze at all. If these sites were representative of interior Alaska, and if the trend continued—both significant *ifs*, Welch acknowledged—then the processes that decompose organic material into carbon dioxide and methane would continue year-round, drastically increasing greenhouse gas emissions.

Welch had asked Kenji for an interview, but Kenji had declined because he disapproved of the story. "Last two years really warm," he acknowledged. "Kinda unusually warm. But many journalists say this is the first time. Never happen in the Arctic, or something like that. Actually, it isn't." He said he thought that the phenomenon in question might well happen in some places once every several decades, and that a few cold years would reverse the trend. "The permafrost damaged, but usually recover. That's the typical pattern of here."

The primary Alaskan scientist featured in the article was Kenji's close friend and colleague at UAF, Vladimir Romanovsky, who had been studying Alaska's permafrost since leaving the collapsed Soviet Union in 1992. When we'd spoken on the phone, he'd called the previous winter "a principal moment in permafrost history."

"It's not like, 'Oops, we crossed the threshold, and we already there,'" he said, acknowledging that interannual variability often drowns out long-term climate trends, creating a "pretty messy picture." "It could be five years, maybe ten years, depending on how climate will be changing, to really cross this threshold and develop this layer which will not freeze again."

NORTH TO THE FUTURE

In the same phone conversation, I asked Romanovsky how long he thought it had been since interior Alaska's active layer had remained unfrozen through the winter. "In some locations, it may happen several times already," he said. "But other locations, I don't know. Maybe several thousand years ago during the Holocene warm time." (The Holocene warm time, sometimes called the Hypsithermal, was a period of global warmth that occurred between 5,500 and 9,500 years ago, for reasons uncertain.) "I can hear Kenji saying, 'Oh yeah, that happening all the time.' But no! If you saying in general, then yes, it's been happening in some locations. But we have documented sites where I would say this was not happening during the last at least thousand years, and now it happened. And I have some data to support this. At least in my opinion."

It was an opinion—albeit a well-substantiated one—because Romanovsky had only studied the sites in question for a few years. His claims were inferences based on multidecadal records from other locations, as well as paleoclimate reproductions, which scientists render using tree rings, air trapped in ice sheets, and other so-called climate proxies. But when I relayed Romanovsky's thousand-year estimate, Kenji balked: "I don't believe that at all. Vladimir not measure thousand years, right?" He pointed out that the only borehole in the world that had been monitored for more than thirty consecutive years was on the border of a spruce forest and a birch forest, which he believed might produce an unrepresentative warming effect on the soil. "Could be very tricky," he said. "Just ten meter away from that site, could be totally different."

At bottom, Kenji did not believe that scientists had enough bulletproof permafrost data from the past to make strong claims about the future. ("Humans don't know much," he said several times.) He seemed especially wary of interpretations that might attract more outsiders to his discipline or give warrant to modelers. ("If you are a city boy, stay in lab, don't go outside, we don't respect you.") Now Kenji made clear that our *nomunication* was over. "Only maybe ten more years I am active, and that climate talk going to waste my time," he concluded. "History will tell everything."

Kenji had forgotten more about permafrost than I would ever know, but I didn't find this last point satisfying. The obvious problem was that by the time history told everything, it would be too late to change course. The early 2000s warnings now seemed, if anything, conservative: The current consensus was that roughly 1.5 trillion tons of carbon—twice as much as exists in the atmosphere—was frozen in permafrost, which in Alaska had been warming at a rate consistent with the most dire climate scenario published in the first IPCC reports. I had seen credible estimates from respected scientists that something on the order of 200–300 billion tons—more than the US and China's total historical emissions, combined—might be vulnerable, and that in a few decades, permafrost might be a larger greenhouse gas emitter than any single country. These were crude projections riddled with uncertainties, yet their implications seemed too large to ignore. And it was my generation, after all, that would bear the cost.

There was a slight chill between me and Kenji as we went to bed that night, and the following evening he departed for Chile. Before walking out the door, he gave me a hug.

"This will be a great time for you," he said. "You will learn a lot."

The temperature, when his truck disappeared into the woods, was −42°F.

9

The night Kenji left it took me hours to fall asleep. Each crackle of the fire seemed laden with meaning. I imagined waking to a burning cabin, or a cabin so cold that I could not build a fire before freezing. I saw a winter bear snapping open the door and eating all my food, then turning to me. These were the fears of a cheechako, but that's exactly what I was, and now I knew it.

Eventually the sunless morning came, and the cabin was still intact, and the bears, as far as I could tell, had remained snoozing in their dens. The fire had gone out; though I'd woken up in the night to add fresh logs, the air was now so cold that the flames had mowed through them in a couple of hours. But it took me only a few minutes, using Kenji's technique, to coax a new flame to life, and soon the front window glowed bright orange, filling me with a deep sense of sturdiness. I ran my hands through the thin, vaguely pubic beard that was overtaking my face, which I knew looked terrible but could not help feeling proud of. *I'm doing it*, I thought, grinning.

That first morning I remained tentative around Shiro. I had watched him devour a dried pig's ear in a few snaps of his great teeth. But when I released him from the pen, he wriggled around me, blocked my path, and rolled over to expose his mountainous, dreadlocked belly. As I ran my hands through his fur, he lay perfectly still, lips closed—and even after I rose and continued down to the reindeer pens, he remained there for a few moments in wishful suspense.

But like a fifth grader working a substitute teacher, Shiro seized on my impostor status and refused to eat his kibble; apparently he was holding out for meat. That night, he woke me with his barking—a sign, Kenji had warned, that predators might be about. I jumped out of bed, grabbed my coat and headlamp, and snuck down to the pens. What I would do if I actually came upon a pack of wolves was unclear, but when I arrived there was no one but Shiro, silently wagging his tail at me with bright, innocent eyes. "The dog who barked wolf," I cooed to him the following morning. He barked again the next night, and by the third I ignored him.

Whether the cold affected the animals' moods, I couldn't say, but the reindeer grew less predictable, too. Some mornings they swarmed to their food greedily as usual, but others they now lingered in the woods. I never saw them move their food bowls, yet I began finding the farthest pen's in different places. On the third day, a scuffle broke out between a cow and a bull that had lost his antlers, during which the cow tore down some old chain-link fencing within her pen. The next afternoon, she got her antlers caught in the torn fence and began thrashing about like a hooked fish.

By the time I admitted to myself that I would have to approach the panicked animal, who was about half again my size, she had tired somewhat, and her writhing had ebbed to intermittent bursts. From a few feet away I saw the crux of the tangle: To free the cow, I would have to bring her closer to the fence, something she was plainly loath to do.

NORTH TO THE FUTURE

Removing my gloves for dexterity, I inched toward the cow and, as soon as her present fit subsided, took hold of her antlers, gently pulling her toward me. She allowed this for a few seconds, and I began to work on the tangle—but then, eyes mad with terror, she ripped loose with a violent jerk; her antlers shot up toward my face and I leaped backward. The outermost tine missed my right eye by inches.

Soon she calmed down again, and several fits later, I managed to slip the fencing free of the last tine; the cow ambled away, bucking appreciatively. But by then my hands had been exposed to the sub −40°F air and heat-sucking metal for several minutes and were barely working. I'd meant to prune the fence to prevent the same thing from happening again; instead I dropped the wire cutters in the snow and ran back to the warmth of the cabin. My fingers stung as they thawed.

* * *

The first half of my time alone coincided with one of the colder snaps to descend on Alaska in recent years. By historical standards, it was not unusual. "No one in Fairbanks 1975 would have been oohing and ahhing about whether it got to 40 below," said Rick Thoman to the *Fairbanks Daily News-Miner* on my second day alone. In Fairbanks 1975, five consecutive January days plunged to −55°F or lower. But the temperature had not reached −40°F in nearly two years, and the local media filled with images of bottomed-out thermometers and lavishly icicled beards. On the third day, newspapers reported on a moose that wandered into an Anchorage hospital, nibbled on some lobby plants, and drifted back out thanks to a door "stuck open because of extreme cold."

As the temperature dropped, the physics of life changed. "Forty below world not the same like other world," Kenji had warned, because at forty below, practically anything—skis, door handles, the plastic casings of generator cables—could break. The air no longer felt like air; thick and heavy, it stung with a sharpness that never let you forget it. Ice crystals formed like cobwebs in the corners of Kenji's triple-pane windows, which he'd warned me not to open, or they would remain open until spring. The atmosphere inverted, trapping a cloud of smog over Fairbanks that glowed in the night like a spectral blanket. By the evening of my fifth day alone, when the temperature bottomed out at −49°F, the air was so dry and still that sounds I'd overlooked a week before became impossible to ignore: the guttural purr of the water reaching a boil; the *plink, plink, plink* of the stovepipe expanding, like drippings on a tin roof; the silky *whir* of hot air filtering through the stove.

Aside from feeding and monitoring the animals, a job which generally took only fifteen minutes each morning and evening, my responsibilities were few, and I spent nearly all of my free hours in the cabin doing research. I felt as though Kenji had left me a riddle, the stakes of which seemed heightened by the cold: *How, exactly, did living this way inform his scientific views? Was there something the other scientists I'd spoken to were missing?*

I was unsure how Kenji would have me go about addressing these questions. The reindeer weren't talking, and staring out at the darkness didn't seem promising. But I got better broadband service from his hilltop than I did from my parents' home in LA, so I began calling his colleagues and asking about the doubts Kenji had raised—the scarcity of study sites, the lack of long-term data, the oversimplicity of models. They all acknowledged the uncertainties inherent in studying global climate, and some grumbled about institutional politics or shoddy papers, but none were quite so skeptical of the new scientific paradigm.

"Scaling's a big deal," said Ken Tape, another UAF field scientist who spends weeks at a time trekking through the Brooks Range for his research. "Okay, so you found out an answer for an area that's the size of my office. Well, you can't just call that 'the Arctic.' Likewise, a modeler who lives in New Jersey still needs to be informed by these fine-scale measurements. For this to really work, all those parts have to be talking to each other routinely. Usually, they are. On the whole, I think we're doing a pretty good job. Not everyone can be strutting across the landscape. It's going to take all these different perspectives."

"There's always more to know, but there's a lot that we *do* know already," said expert Ted Schuur, who leads a research group called the Permafrost Carbon Network. Schuur and his colleagues have found that by the middle of this century, thawing permafrost is likely to emit between 400 million and 1.5 billion tons of carbon per year, producing a warming effect roughly equivalent to the 2023 emissions of Brazil or the US, respectively. "We have reliable ways of aggregating uncertainties," he continued. "Reality is not going to be five times our worst-case scenarios. And it won't be one-fifth of our best-case ones, either."

But Kenji was not trying to forecast climate change; he was trying to archive pure facts. ("Hundred years from today my data still valuable, because is truth, *only* truth," he'd told me. "I don't speculate.") He was seeking to observe the land directly, without excessive technological clutter. Still, I wasn't sure what Kenji expected me to learn alone on his homestead, or how he would have me go about learning it. The information I wanted was not to be found in his cabin, but it was all, always, only a few clicks away. So I downloaded papers and books on permafrost, meteorology, and reindeer herding. I called experts and coded my notes, drifting free of the cabin for hours at a time.

On my third afternoon alone, I got so caught up reading about meteorology's role in the standardization of time—a nineteenth-century development that helped untether human activity from the rhythms of the land and sun—that I forgot to tend the fire. Realizing that I was cold, I rose to throw a few logs in the stove and noticed that it was already dark out. The daylight was brutally short—basically just one long sunrise and sunset—and I had missed it.

*   *   *

On the night of my fifth day alone, the temperature bottomed out, and then began to rise. Within two days it hovered just below zero.

I no longer had to feed the stove every half hour, and the air outside didn't sting. There were no more scuffles between the reindeer; Shiro broke his hunger strike. Before leaving, Kenji had several times described my job in fairly harrowing terms, no doubt trying to stoke my frontiersman fantasies: "So you feel like you can live in forty below, middle of forest, alone?" he'd asked once. "If problem, if you no die, later, good story!" he'd said another time.

But my tasks around the homestead began to feel habitual; mostly they involved sitting around. Before Kenji had left, we'd made a run into town to buy groceries, and the cabin was stocked with so much food that I could have made sandwiches for the reindeer. I had more wood than Paul Bunyan. Even if I managed to royally screw up, my broadband service was plenty good enough to call the paramedics.

Gradually my position began to feel less extreme than I'd imagined, and an uneasiness set in. Maybe Kenji wasn't trying to teach me anything; perhaps I was just some chump who'd agreed to housesit for free. Maybe eleven days was not long enough to crack the code of Kenji's life, and maybe there was no code to crack. I had read dozens of articles and taken many pages of notes; I had gotten cold, learned to build a fire, and wrangled a reindeer; but I didn't feel changed—just lonely. And as my sense of purpose ebbed, I was confronted, for the first time, by the sheer weight of the silence, which was not actually quite silent: There was always the sound of the fire, licking the logs like a dog devouring a shank of meat. I felt the darkness, which was not entirely dark: Even outside, there was always a soft glow of stars on the snow.

It didn't occur to me that this might be where the true learning began: When the story I'd been telling myself fell apart. When reality broke me down enough to see what was actually there. I knew only that I was uncomfortable and didn't want to be. In this frame of mind, the internet began to serve a somewhat different purpose: It was no longer just a research tool, but an escape portal.

I continued to research, but my research was losing focus. I kept downloading articles, then promptly identifying flaws that excused me from reading them. There was always another article, and this promise kept me skimming along like a water bug. Soon I slipped into my old digital habits—disappearing down YouTube rabbit holes, mining Instagram's "explore" page, letting information wash over me like a stimulant. The sounds of the fire disappeared behind 128-beat-per-minute Spotify "work" playlists.

The early champions of the internet celebrated the technology's potential to liberate us from our surroundings and even our materiality. "Ours is a world that is both everywhere and nowhere, but it is not where bodies live," declared the cyberanarchist John Perry Barlow in 1996. As a tool, I was learning, this everywhereness had become indispensable, but as a way of life it can be deadening. Being in a particular body, in a particular place, imbues our particular surroundings with significance. It solicits the kind of intense, often uncomfortable engagement that deepens our understanding of the world and our place within it. Everywhereness threatens to dissolve this significance, because when everything is available, everything can seem trivial. When nothing is far, nothing is near. We are always an arm's reach away from opting out.

Alone in Kenji's cabin, my thoughts grew restless, and I fell into a kind of hyperactive ennui. It seemed pathetic that I had traveled to one of the wildest environments on Earth only to spend my time toggling between newsfeeds. But even when I considered turning off my devices for bits of time, it felt contrived: *Why pretend? Who was I kidding?*

So it was that in one of the more extraordinary places on Earth, I managed to pass some of the more banal hours of my life.

* * *

There were wonderful moments, though. I hardly recognized them at the time; I was too fixated on my own shortcomings. But seven months later, by the end of my first Brooks Range expedition with Roman, I had been offline long enough to know how it felt to be present. I thought I was finally beginning to understand, in some provisional sense, what Kenji might have been trying to preserve—or at least what I hoped to preserve myself. And as I sat in my hostel waiting for the smoke to clear, certain memories from my time alone at Kenji's surfaced.

There was a clear pattern to the moments I remembered: They had nearly all taken place during the hours I'd spent outside, exploring Kenji's trails. After lunch each day, I would don the rabbit fur gloves Kenji had given me and head outside to ski with Shiro. I brought my phone on these trips but couldn't use the internet; broadband was spotty on the trails, and it was too cold to take my gloves off for long, and my phone's battery drained rapidly unless I kept it under several jackets.

Initially I skied to let Shiro vent his boundless energy and to take the edge off my own endorphin withdrawal. In light of the cold weather, Kenji had not asked me to patrol his property, and on the first day I kept to the gated clearing. But I was curious about the trails, and on the second day Shiro forced my hand. After our first lap, as we reached the far end of the clearing, he burrowed under the fence and disappeared into the forest. I hesitated for a moment, and then—with a mixture of apprehension and relief—followed.

NORTH TO THE FUTURE

We didn't make it far that first outing. As I began my second lap of a short loop, I came upon a set of large tracks I hadn't noticed there fifteen minutes before and beat a hasty retreat. But in the days that followed, I let Shiro lead me deeper into Kenji's maze. Now and then I pulled out my phone to snap pictures of prints, which back at the cabin I would use to guide internet searches. In this way I came to recognize the furtive paddings of red foxes, the meandering gouges of moose, and the well-traveled thoroughfares of snowshoe hares. I learned that the tawny squirrels clattering the frozen trees were red squirrels, and that the plump songbirds twittering from boughs were black-capped chickadees—their toothpick legs kept warm thanks to dense meshes of veins called *rete mirabile*, Latin for "wonderful net." I couldn't seem to wrap my head around this fact—that a creature so fragile-looking, so *cute*, could survive all winter in temperatures that sent me inside within ninety minutes.

On the seventh day, as Shiro and I crested the pingo, a yellow-eyed canid emerged from the trees. Whether it was a wolf or a coyote, I couldn't tell, but Shiro spotted it the same moment I did and took off in the creature's direction. For the next fifteen minutes, as I called to Shiro in vain, my mind filled with grisly images of torn fur and blood-soaked snow—the fate of many dogs in rural Alaska. But eventually Shiro emerged from the trees, panting and wagging his tail.

The next day we returned to the spot and followed the tracks of the animal, a lone coyote who had apparently circled the property several times, often trotting slowly, other times covering a dozen feet per stride. In the days that followed, we explored Kenji's entire trail system an hour and a half at a time in pursuit of the elusive creature. When we paused to inspect tracks, I could sometimes hear the plasticky crunch of a reindeer's footsteps from hundreds of yards away. I could hear the willow *whoosh* of a distant raven's wingbeats as if it were flying by my face. The perfectly still air was occasionally ripped by a shrieking howl, as though the atmosphere were tearing open. Whether the noise was caused by the wind-shear of far-off cars, or winds on distant mountains, or something else entirely, I never figured out.

As the days grew longer, the midday sun—still low on the horizon—enriched the land and sky to Egyptian golds and blues. Through the crisp air I could see Denali 170 miles away, but its visage was often warped by Fata Morgana, a bizarre variety of mirage named after King Arthur's shifty sister, the sorceress Morgan le Fay. The inverted atmosphere blurred vertical lines and flattened horizontal ones, melting the Alaska Range into the layered mesas of southern Utah. Sixty-million-year-old land masses liquefied and reformed before my eyes. The air was cold enough to kill, yet the twilight dripped lava-like across the mountains, candescing the crystallized scape in colors warmer than any I had seen in Los Angeles.

Now and then, I felt what Martin Heidegger called the "the pain of the nearness of the remote." I saw just enough to sense that I saw almost nothing—that each set of tracks leading into the woods held only a trace of the creature who left it. That beyond each trail lay a vibrant world that I lacked the skill and strength to explore. These moments filled me with fear, but also longing. It seemed to me that a different, better being could have found room in itself for far more of that wildness, and I wanted badly to become that being.

But this fear and longing also illuminated what was nearest of all: the feeling of the cold air in my nose; Shiro's grace as he plunged through the snow, dark eyes gleaming with joy; the ever-changing colors of the land, which some wordless part of me knew I would never manage to capture. I still don't know for certain what Kenji hoped I would find during my time alone, or if he really cared. But years later, these are the moments that emerge vividly from the dark sweep of lost time: The moments I was really, finally, there.

10

The morning after my twelfth night alone, Kenji returned, and that afternoon he took me on his first reindeer ride of the winter. It was unusual for him to bring someone else along, because the animals are so skittish. But after two and a half weeks they had grown comfortable enough with me that he was willing to try.

We walked down to the pen, where Kenji began wrangling the calmest of the animals, a castrated male who was typically the first to approach whenever I arrived with food. He was still a nervous wreck. It took Kenji several minutes to fit a halter over the beast's head and then drag him over to a corral, where—as the animal paced in evasive circles—Kenji proceeded to affix a harness one buckle at a time. It was a tense affair: Kenji cooing, the animal trembling on the edge of panic. Even once the harness was firmly attached, Kenji kept the deer on a short leash, because if he got himself tangled a panic attack would ensue. Then he might be unapproachable for weeks.

Out on the driveway, Kenji had left a dog sledge waiting—a lightweight metal frame with a basket roughly the size of a yoga mat. As he attached the reindeer harness to the sledge, he had me sit in the basket; then he began running to scare the reindeer. It worked. The reindeer took off at a trot, the sledge lurched forward, and Kenji hopped onto its runners. But the ride was short-lived. After perhaps fifty sluggish yards, the reindeer pulled over to sniff a tree. Kenji groaned.

Suddenly the animal took off again, and we whiplashed back into motion. In the process, the rope somehow got wrapped around the reindeer's hindleg, and he moved slowly now, hardly faster than I might have jogged. Nothing about the outfit seemed remotely promising. We were at the whim of a skittish, untrained creature who, after a winter in a pen, was not even particularly fit. Near the end of the driveway, the reindeer stalled again, and Kenji decided to call it a day. If we kept going, the animal might panic, he said; better to quit while we were ahead.

As we walked the reindeer back to its pen, I found myself wondering what the hell Kenji was doing with his talents, and why he insisted upon doing it mostly alone. People do manage to run reindeer; the Sámi of northern Europe, for example. But they do so in communities called *siiddat*. They build a culture around it, passing on knowledge from generation to generation, raising thousands of animals over lifetimes and then picking the very best ones for the purpose. Some things in life—some of the most beautiful and impressive, for that matter—are like this: They take a village, and sometimes far more.

I was grateful that there were people like Kenji out there, who questioned the assumptions most of us accepted so lightly. He had earned his views the hard way and shared them only when asked. But by the end of my stay at the homestead, I felt hungry for solutions. Not even Kenji disagreed that climate change was likely to thaw a great deal of carbon-packed permafrost at some point, so the question remained: How to produce the best possible information in order to avoid the worst outcomes?

Direct observation was clearly necessary, but it was also insufficient.

("The very notion of 'climate change,'" observed the philosopher of technology Benjamin Bratton, "comes from the sensing, modeling, and calculation of the measurable change in a planet in ways well beyond direct human reckoning.") Our only path forward was to work together with the data and tools available, and none of Kenji's points had convinced me that this was quite so unworkable as he seemed to believe. If anything, I left his homestead more curious about the techniques scientists were developing to study the land at scale, and it was this curiosity that led me to the glaciologist Matt Nolan.

But the following summer, in the days after my first expedition with Roman, I realized that Kenji was, in at least several key respects, profoundly correct: The Alaskan wilderness remains unfathomably complex. Viewing the world through a screen can produce an undue sense of certainty and self-importance. There is a risk, as we grow more dissociated from the workings of nature, that we fall prey to simplistic narratives about its fate. And in pursuit of planetary knowledge, we often sacrifice direct sensory engagement with our surroundings. It was this kind of awareness that had brought the natural world to life for me in the Brooks Range, pulling me out of my own head and into the realm of fish and wolves and trees. Revealing to me the true stakes of Roman's research, and the richness of the world beyond my screen.

During my winter at Kenji's, I had become so fixated on the effects of climate change that I'd spent almost no time observing the place itself. The science, and the tools I'd used to approach it, had consumed virtually all of my attention—and as I sat in my hostel room the following summer, I resolved to not let that happen again.

* * *

On my final evening at Kenji's, Vladimir Romanovsky and his wife, Noel, came over for dinner. A Fairbanks biologist named Skip Walker was running late, and the Romanovskys worried that the downed moose they'd passed on the highway might be responsible. But soon Walker arrived, and the drinking began.

After dinner, we donned jackets and walked down the hill to Kenji's banya, a Russian woodburning steam bath that looks from the outside like a small log cabin. According to Kenji, Russian tradition dictates that a pair go in and out of the steam room until they've finished an entire bottle of vodka. Most Sundays he and Vladimir observe the tradition. But tonight, we sampled different liquors and snacked on grapes and shrimp, the scientists swapping stories from their travels. And as we sat there wrapped in sheets, sharing laughter, feeling our skin drip and the alcohol metabolize, I found it deeply reassuring to think that people who disagreed so vehemently about weighty matters could still get together in this way. And this reassurance surprised me: Shouldn't this have been obvious?

Late in the evening, conversation turned to a plum tree Kenji had recently planted in the clearing of his homestead. Near the tree he had built a snorkel-stove hot tub, which—once filled—he would have to heat constantly to keep from freezing. The tree was young and would take time to grow, but Kenji told us he looked forward to the early spring day when it would finally blossom. Then he would sit in his hot tub and look out at the flowering tree, contemplating its beauty.

Walker, who understood the Fairbanks climate as well as anyone, laughed. "All that work for a few hours, if you're lucky!"

Kenji just smiled.

PART III

MAPS OF McCALL GLACIER

A man who keeps company with glaciers comes to feel tolerably insignificant by and by.
—Mark Twain, *A Tramp Abroad*

It is not down on any map; true places never are.
—Herman Melville, *Moby Dick*

1

July 2019

Days passed in Fairbanks; the fires continued. Entire forests were incinerated, thickening the atmosphere to a dim, ashy glow. Initially my hostel seemed a welcome reprieve from tent life, but soon the thrill of hot showers and Häagen-Dazs wore off, six weeks' worth of missed messages began screaming for reply, and I felt my mind closing in on itself again, much as it had the winter before.

I had stayed in Fairbanks, rather than continue on to Anchorage with Russell and Julia and Duncan, because I hoped to return to the Brooks Range, this time with the glaciologist Matt Nolan. Alaska's glaciers currently account for roughly one-quarter of global ice loss, and Matt was the only scientist conducting fieldwork on Arctic Alaska's glaciers, which are scientifically fascinating but extremely remote. To accomplish this, he had developed a technology that was making waves in Alaska—a tool that allowed him to survey entire landscapes with an unprecedented combination of speed, precision, and economy. I hoped to learn from him what it might look like to combine direct observation and remote calculation, the embodied and the digital. And more to the point, I wanted to go back out.

The weather was not cooperating, though. We planned to leave on my seventh day in Fairbanks, but that morning I awoke at four thirty a.m. only to find a text from Matt: a storm had blighted the forecast. So more days passed inside, online, and I grew frustrated by how easily I slipped into my old habits. The Brooks Range had apparently not cured me of the impulse to binge YouTube videos or check my phone every few minutes for vapid updates about political mayhem I couldn't control. The presence of mind Roman had coaxed out of me proved fleeting and dependent on upkeep. That larger world of interspecies connections began to feel like a separate dimension, which I could access only by going back to the wilderness. I perused the internet for dry cabin rentals around Fairbanks, toying with the idea of sticking around when the summer ended.

After a few more days Matt identified another possible weather window, opening on July 23, my twelfth day in town. I didn't sleep much the night before. My bags were packed, and I had nothing left to do except brush my teeth and guzzle some coffee, but I was anxious about missing our rendezvous. I was scared, moreover, of the part of me that wanted to miss it—the part that was content to keep scrolling.

*   *   *

Matt arrived outside my hostel at five thirty the next morning, driving an old blue pickup truck with a cracked windshield. The sky was orange. I tried not to breathe too deeply as I loaded my pack and climbed into the cab.

"How's the weather looking?" I asked as we began toward the small-plane airfield.

"The weather's fine," said Matt. "It's the smoke that'll fuck us."

NORTH TO THE FUTURE

Along the empty roads, warm winds bowed the spruce and jangled the aspens, and there was no telling where they would blow the fires next. But there was no telling either when the next window of decent flying weather would open, and after nearly two weeks inside, we were both willing to roll the dice.

The airport materialized in the hazy morning light, and we drove through a back gate onto the tarmac. After passing several rows of colorful bush planes, we pulled up next to an old blue and white Cessna 206, a single-engine six-seater. Matt pulled his key out of the car's ignition before we had come to a full stop, and started hurriedly transferring supplies from his truck to the plane.

He wore a hoodie that said FLY IT LIKE YOU STOL IT—STOL being an acronym for *short takeoff and landing*: a point of pride in the bush, where pilots often have to maneuver on a dime. Tall and lanky, with a ponytail and a thick salt-and-pepper goatee, Matt was a striking figure, but he did not possess Roman's or Kenji's Olympic grace. He looked less like an explorer than a hard-living coder. But he moved carefully, placing items like puzzle pieces to balance the plane's weight. Soon the plane was loaded, and he told me to get in.

"Be careful about what you grab," he added. "This plane is older than I am."

I did not relish this fact; Matt had been born in 1966, before cars had airbags. But it didn't surprise me, either. When I'd met him here at the airstrip the summer before, he'd been talking with a mechanic about fixing the plane's propeller, chipped by loose gravel at an unpaved airstrip.

"Is this repair number four or five in the past two months?" the mechanic had asked.

"That's the price of playing poker," said Matt.

"Pretty soon, there won't be any entries in your log going outbound," replied the mechanic. He seemed to be teasing Matt but also warning him. Another bush pilot had advised against flying with Matt. "*I* wouldn't get into a plane with that guy," he'd said.

We hoisted ourselves into a cramped cockpit and clipped into shoulder harnesses. Matt showed me how to fasten my helmet and then pointed out several of the plane's safety features: radios, fire extinguishers, and an emergency locator transmitter, which would broadcast our location if we crashed. Matt had been in several plane crashes before, though none as a pilot, and we each carried a jacket, a fire starter, and a satellite device in case it happened again.

Matt turned the key, and the engine hummed to life. We jounced onto the runway.

"Cleared for takeoff two-right, Stationair One Delta Charlie," he radioed to the air traffic controller. The plane rumbled forward, and within a couple hundred feet we were in the air.

We climbed north over the White Mountains, their valleys shrouded in smoky haze. Rust-colored sunlight pierced the haze and glanced off lakes and rivers, mirroring our movement over the landscape like an endless fuse. Matt opened the glove compartment and pointed to a crumpled, water-stained booklet. The words on the cover were mostly illegible, the staples rusted. "That's the plane's owner manual," he said. Beside the manual was a handgun. I asked what it was for. Matt shook his head, as if the question were too naïve to merit an answer.

"For me, if I ask too many stupid questions?" I asked. He smiled.

As we traveled north, the smoke grew thicker. Between the White Mountains and the Brooks Range lie the Yukon Flats, which are as flat as Illinois—but for all we could see, they might have been the Rockies.

"I can't even see the ground," said Matt into the radio.

"You should break out of it," said a nearby pilot.

"I hope so," said Matt. "I'm hoping to land on a glacier."

2

had first heard about Matt within days of arriving in Alaska the previous summer. He'd just produced a map of Denali that was orders of magnitude more precise than any before, and in a state that worships mountains, anything to do with the crown jewel causes a stir. The US Geological Survey's latest map of Alaska, completed in 2013 after several years of airborne radar surveys, found the mountain's summit to be seventy-three feet lower than its true value (determined by ground GPS). It had fifty-foot resolution, which meant that a forty-nine-foot cliff could hide. Matt's map—which he'd made in one day of flying and for a small fraction of the prorated cost—achieved six-inch resolution. He'd pegged the summit altitude exactly.

A few days after hearing about this, I emailed Matt, flew from Anchorage to Fairbanks, and met him at the airport for his propeller repair. A few days later I biked to his home on the outskirts of town. It was mid-July; sunny skies lit up all of the interior, and fireweed popped like fuchsia fireworks along the roads. After a couple of miles on paved roads and a couple more on dirt ones, I arrived at a bizarre, rambling structure: What began as a one-story building on one side rose to three stories on the other, atop which perched a balcony with wooden battlements. Matt calls the building his castle.

"The kernel that makes Fairbanks different, I think, is the lack of a building code," he explained after showing me inside. "You can live in a tent if you want. You can live in a palace if you want. You get to choose how you spend your money or whether you even need money. You can live for nothing. A fifty-pound bag of rice and a fifty-pound bag of beans will cost you fifty bucks, right? And you can live on that all summer. Whereas the city puts you on the hamster wheel. You can't stop running or you'll fly off."

Matt grew up in Middlesex County, New Jersey, the most densely populated state in the union. After completing an engineering degree at Carnegie Mellon in 1988, he couldn't bring himself to take an office job, so he and three friends drove to Alaska to find work. After a few months in a fish cannery, Matt's friends left, but he stayed. He worked in IT and lived out of his car for a year before taking out a loan to begin a master's degree in Arctic engineering at the University of Alaska Anchorage; soon thereafter he began his PhD in geophysics at the University of Alaska Fairbanks, and in 1999 became a research professor.

"When you start to make lifestyle cookie-cutter," Matt continued, "you make thought follow that lifestyle. You have to conform. And it's not like some evil genius is doing it to you. It's just—you buy your food at Fred Meyer, and you buy your stuff at Home Depot, and it's easy. You're in the mainstream. Everything is…no problem. But it changes the culture. It changes people's thoughts."

The focus of Matt's research was a glacier in the eastern Brooks Range named McCall, which he believed could provide insight into the future of the Greenland and Antarctic ice sheets. No one knows how fast the world's ice will flow into the ocean; credible estimates of sea level rise still range from eleven inches to six and a half feet by 2100, and the difference is a matter of life and death for low-lying places such as New Orleans, Bangkok, and much of Bangladesh. But measuring glaciers is arduous, time-consuming work, and in 2011, Matt began developing a mapping technology that would radically accelerate the process. This project eventually drove a wedge between him and the university, and in 2016, Matt left his professorship to conduct research independently.

"He has a fierce sense of independence," said the dean of UAF's College of Engineering, Bill Schnabel, who described his own position as "the person who had to define the lines that [Matt] didn't want to stay within." "There's kind of a big difference between how he wants to do things and the way most science is done these days."

The flashpoint came not over the mapping technology itself, but over whether Matt could fly his plane to research sites—a prerequisite for using the tool. The university refused to reimburse his expenses for doing so, despite the fact that it was far cheaper than chartering a helicopter or third-party pilot, as he'd always done in the past. "They call it 'risk management,' but it's really liability management," said Matt. "They're trying to throttle you back all the time."

Of course, landing a single-engine plane on a glacier hundreds of miles from the nearest city is more than a liability concern. It's downright dangerous—even for a seasoned bush pilot, which Matt was not. He began learning to fly in 2012, at the age of forty-six, after a plane crash during fieldwork left him (a passenger) with a broken rib and a reason to take yet another matter into his own hands. Within two years he was landing on the glacier alone. In the middle of one of his book-length blog posts, written after his first solo flight to McCall, Matt shared some thoughts on risk management:

> If you're coloring completely between the lines, the fact that there are lines means that someone else has

determined that you can be 2–3x outside the lines and
still have a strong margin of safety. . .

"I was worried that I would read in the news that he had crashed and died," said Bernhard Rabus, a colleague who himself had spent years doing fieldwork at McCall in the 1990s. Ken Tape, another colleague, said that most people would call Matt's solo fieldwork "crazy," but that he thought "bold" was more accurate. "I give him the benefit of the doubt because he's a very bright guy. He's a very bright guy who's taking, I think, very calculated risks."

Indeed, Matt had a good reason for flying. Leading me through his castle, he showed me up a ladderlike staircase to his office, a spare den painted gray. There he fired up a computer and four monitors arranged on an L-shaped desk. On the walls hung several Shakespearean insults:

I love thee not, therefore pursue me not.
What great ones do the less will prattle of.

He pulled up a series of bird's-eye photographs of the Arctic landscape that he'd taken—one every few seconds—through a port in the belly of his plane. "On a really productive day, I'll take about fifteen thousand," he said. He was now checking the photos for quality, and next he would pair them with GPS points—taken at the same time as the photos—and then run billions of automated equations to turn this information into a highly detailed, 3D digital map. Think Google Earth, but orders of magnitude more exact: a virtual Alaska. The land inside of a computer.

Matt's maps were not, in either a technical or a historical sense, truly novel. The technique he used, called photogrammetry, had existed in some primitive form since the 1850s, and is essentially a process of triangulation: by taking several photos of the same thing from different angles, one can apply formulas to produce a 3D model. Photogrammetry is one of a suite of techniques (using radar, lasers, and other airborne instruments) scientists are now using to render ever-more precise representations of the Earth, and this revolution in so-called remote sensing is itself merely the latest chapter in humankind's centuries-old campaign to draw nature into the realm of rational understanding. "The fundamental event of the modern age is the conquest of the world as picture," wrote Martin Heidegger in 1938, echoing the modern credo that to represent is to know, and to know is to control.

But Matt had harnessed modern processing power, a penchant for tinkering, and one key (but proprietary) technical insight—in how to more precisely align his GPS information with his photographs—to bring the method into a new age.

"What Matt did to photogrammetry," said Rabus, now a professor at Simon Fraser University who specializes in remote sensing, "is something that seemed, and was, quite obvious. But he was the first to see the obvious. And that's why his system is far more accurate than anybody else's."

Matt's maps were far sharper than satellite imagery, far faster to produce than ground-based mapping, and far cheaper than lidar—a method of scanning landscapes by laser. "He found a sweet spot," explained Rabus. And this sweet spot allowed for a host of novel applications. By making a map of the same place in successive years, he could compare the two to find needles in haystacks. He could singlehandedly detect changes in glacier volume, coastal erosion, stream channel migration, permafrost collapse, polar bear denning, and all manner of other subtle but significant changes in the land.

As my castle tour neared its end, I asked Matt to show me the

Denali map. My reasons were personal. A few weeks after he'd made it, I had climbed the north side of the mountain, as arguably the least qualified member of a thirteen-person, twenty-six-day expedition. The mountain had terrified me. During our fifth day on the glacier, one of our leaders broke through a snow-covered crevasse and fell thirty feet, breaking two ribs. The next three weeks brought an avalanche, gale-force winds, and a house-sized block of ice that collapsed ten minutes after we passed under it. The weather whipped from searing sunshine to driving snow to whiteout fog with such schizoid frenzy that I had the distinct feeling, as we crept up the mountain, of scaling the spine of some asteroidal god. At last a clear window of weather permitted us to scurry to the wind-blasted summit, but what I felt, upon returning to the land of the living, was not so much triumph as relief, even shame. I had not conquered Denali: I had merely survived it.

Now, sprawled across Matt's monitors, was that same mountain—each granite outcrop, yawning crevasse, and fluted ridgeline. Every avalanche fan, merengue snow pillow, and wavelike zastruga rippling the surface of the snow. It was all reproduced with uncanny precision. My awe for the mountain became awe for Matt: this being who, through sheer ingenuity, had transformed the largest aboveground mountain in the world into a toothless play of lights on a screen.

Matt zoomed in on the behemoth's surface. With each flick of his finger, he moved the great mountain. With every double click, he revealed its closely guarded secrets.

"If there had been climbers out there," he said, "you'd see them."

3

Two hours after takeoff, somewhere near the southern foothills of the Brooks Range, our plane busted free of the smoke. Haze had begun fingering into the lower valleys, and the horizon behind us was shrouded in billowing yellow, but the sky before us was blue. Around nine thirty a.m., I dimly registered that we were crossing roughly the midway point of the route I had traveled with Roman a month before, near the valley of the howling wolves. Then the plane rose above the greenery and into a wind-scoured realm of rock and ice: We were approaching the Romanzof Mountains.

A subrange roughly sixty-five miles long, the Romanzofs are the tallest mountains in the Brooks Range, or anywhere else in the North American Arctic. (No one knew this fact for certain until Matt mapped them, correcting USGS figures from the 1950s.) They also hold the largest glaciers in the American Arctic; hundreds churn through the granite valleys. All of this ice lies more than a hundred miles east of the Haul Road, in the heart of the 19-million-acre Arctic National Wildlife Refuge.

"So this is your office," I said into my headset, as we climbed toward the highest mountains on the horizon.

"In a way," replied Matt, looking up from the gauges on his dashboard.

The Romanzof glaciers are not large compared to the other tens of thousands of glaciers farther south that cover about 5 percent of Alaska. They are downright puny relative to the Greenland and West Antarctic ice sheets, which will largely determine future sea level rise. The smaller of the two, the Greenland ice sheet, holds almost seven hundred thousand cubic miles of ice, enough to bury all of Texas two and a half miles deep. It releases more water in a day than McCall Glacier has shed over the past half century. But because the Romanzof glaciers are both further along in their degradation and simpler to study, they may offer crucial insight into the future of the great ice sheets.

"The problem with Greenland is that you can't get your arms around it," explained Matt. "You can't get a complete system view, because it's just so fucking massive. But what's happening on McCall now is essentially the same thing that's gonna eventually happen on Greenland. Ice is ice. Glaciers are glaciers. The scales of Greenland and McCall are very different, but the core dynamics are almost identical."

There are plenty of relatively small, cold glaciers in the Arctic; what makes McCall special is its long history of research. In 1957, the US Air Force funded a scientist named Richard Hubley to lead an eighteen-month study on McCall. Military aircraft dropped roughly fifty tons of supplies onto the glacier below, and Hubley's team planted a dense network of stakes in the ice to track its movements, and thereby to calculate changes in its size and rate of downhill flow. In the years that followed, another team studied McCall from 1969 to 1972, and another from 1993 to 2001. By the time Matt revived research in 2003, McCall was one of the longest-studied glaciers anywhere in the Arctic.

NORTH TO THE FUTURE

"It's an amazing wealth of data," said Frank Pattyn, a leading ice sheet modeler who uses the McCall data to better understand Greenland and Antarctica. "It's a unique laboratory. There are not so many glacier studies in the world that are so well-established and have such a neat record."

Matt and Frank hoped to learn from McCall what would happen when glaciers stopped producing any new ice. A healthy glacier has two zones: The upper half, high on the mountain, is the accumulation zone, where snowfall tends to last year-round, compressing the snow beneath it into a more granular snow called firn—and eventually, over the course of decades, into ice. All of this ice is slowly flowing down the mountain into the so-called ablation zone, where snow melts by the end of the summer and the glacier ice is steadily melting. As the climate warms, the accumulation zone on most glaciers is shrinking, and on McCall it had already disappeared entirely. Since Matt had begun research, McCall had become one big ablation zone.

"The same thing will eventually happen in Greenland, and even Antarctica, if the current trends continue," explained Matt. "It may take a while, but it will."

The question, then, was how the end of new ice would affect the rate at which glaciers receded, and Matt was in a unique position to address it. But in 2013, Matt learned that his latest grant proposal had been rejected. Perhaps it was his ornery exchanges with grant administrators and logistical coordinators, who for years had been trying to rein him in. Maybe the competition that year was especially stiff, and his application simply got lost in the shuffle. The truth is that it had always been difficult to secure consistent funding for long-term, expensive baseline studies—hence the stop and start nature of McCall research since 1957. ("It's a monitoring program," said Pattyn. "And that is not so sexy.") Whatever the reason, Matt faced a painful choice.

"Do I organize a giant demobilization project?" he wrote in his field notes, referring to the laborious process of removing all existing equipment from the glacier. "Do I walk away from it, like everyone here before me has done?"

Instead, he made some modifications to his plane. He practiced "making hard banks at low level," "doing stall-horn-screaming-chandelles," and "landing on a dime on a strip no wider than my wheels in a crosswind"—and in August 2014, just nineteen months after his first flight as a student pilot, he landed solo on the glacier. "I've never been a fan of debt, and I realized now that all grants are a form of it," he journaled during that research trip to McCall. "If I'm going to figure something out scientifically, it was going to be from cash, not debt. Whether that was possible to pull off or not I couldn't really say, and still can't, but it's that way or no way."

His new plan was to study McCall alone, and on his own dime—no bureaucratic strings attached. By flying himself to and from the glacier, he figured he could eliminate the exorbitant cost of helicopter transport, along with the expense and logistical headaches of traveling with others. ("I was doing most things on my own anyway," he wrote. "Other people were mostly there to facilitate body retrieval.") His secret weapon was his new mapping technology, which he would use to raise his own research funds by making high-quality maps for scientists, commercial interests, and federal agencies. It would also enable him to collect many of the most important glacier measurements from his plane, with far greater speed and accuracy than was previously possible. As late as the 1990s, calculating the volume of McCall had required skiing around it with a GPS and a bulky instrument called a theodolite, and then extrapolating from individual data points.

"Now," he explained as we neared the mountains, "I can just measure the whole damn thing."

NORTH TO THE FUTURE

* * *

Decades before computers and satellites, years even before the first modern maps of Northern Alaska existed, Ernest Leffingwell set out on the first documented journey into the Romanzofs. It was May 1907, exactly half a century before research began on McCall Glacier, and the northern coast was known to the Western world only by the accounts of a few whalers and ship-bound explorers. To them the interior remained terra incognita. Not even Native Alaskans lived in these mountains. Iñupiaq families had once inhabited the foothills, but they had retreated to the coast in recent years following a precipitous decline in the caribou herds they hunted—and by 1907, according to Leffingwell, the mountains were considered taboo.

Leffingwell was shipwrecked. He had just completed his first objective in Alaska, a perilous scramble over sea ice in search of the mythical Keenan Land, a landmass thought to lie some three hundred miles north of the continent—and upon returning to shore with definitive proof that the myth was, in fact, myth, he found that winter sea ice had wasted the wooden hull of his schooner.[1] Most of his crew had decided to hitch a ride back to civilization on the next passing whaling vessel, but Leffingwell turned his attention inland, to a towering mountain range "so covered with ice and snow as to be conspicuous from all directions." Setting out on dogsled in mid-May with two prospectors (only later would he discover the helpful wisdom of Iñupiaq guides), he traveled some fifty miles inland over the snow-covered tundra to the base of the mountains, and from there he continued on foot.

The journey was arduous and slow. The river he followed into the mountains—known to the Natives as Okpilak, or "no willows"—was

---

[1] This was a common fate for wooden ships in the Arctic. Those forced to spend a winter in one sometimes called the cracking sounds the ice made as it tore apart their homes "the devil's symphony."

the steepest in the region, and so swift that "one frequently hears the sound of rolling boulders." He lugged a variety of heavy scientific instruments, including a full-sized plane table for cartographic measurements, and fed himself by pausing to hunt ptarmigan, caribou, and sheep, the latter of which he reported seeing "as many as 40 or 50 in a day." When at last, after three weeks of grueling travel and a dicey river crossing that swept him off his feet, he reached the head of the river, he was not disappointed: Before him lay the largest glacier in the Romanzofs.

Leffingwell was not easily given to excitement. A stoic man, he stuck to scientific writing and never wrote public-facing accounts of his exploits, which partially explains his relative obscurity in the annals of polar exploration. But in describing Okpilak Glacier for a USGS report, he departed from his usually Spock-like prose to paint a scene that was defined by what he could not measure. He wrote of pure ice that extended "to the limits of vision," rivulets of spring snowmelt that "disappear into crevasses," and four-thousand-foot granite walls so sheer that their tops were "invisible from the valley floor." "The whole area around the head of this glacier is so covered with snow and ice that scarcely a rock can be seen even in the summer," he recalled. The forbidding terrain curtailed his survey, but before leaving, he took several photographs. It was the last time anyone would see America's largest Arctic glacier before it began to rapidly retreat.

For nearly a century, Leffingwell's photos sat in an archive collecting dust. Then in the early 2000s, Ken Tape stumbled across digital scans of several of them online. Matt tracked down the originals and digitally stitched them together into a panorama of the glacier. Here, he realized, was an invaluable time capsule—and if he could find the place from which Leffingwell had taken the photograph, and repeat it, then he could visually capture how the glacier had changed over the preceding century.

NORTH TO THE FUTURE

In 2004, Matt took a day off from McCall research to search Okpilak. Hiking over streams and glacial debris, he eventually crested a rocky knoll and "nearly shouted with joy": before him lay a small rock cairn, which Leffingwell had evidently built to mark the spot. But three years later, when Matt returned to repeat Leffingwell's photo on its centennial anniversary, the cairn was no longer there. The part of the knoll on which the cairn once sat had collapsed; it had been underlain by ice, and the ice had thawed.

Now, with me in the plane, Matt decided to use his mapping technology to repeat the photograph, remotely, before we continued to McCall. "The weather's good, and it's still early," he explained, his gravelly voice crackling over the headset. He adjusted his hand and the plane tacked west.

Around us, just beyond the plexiglass bubble of the cockpit, jagged black peaks and dinosaur-spine ridges carved the sky. Tongues of ice oozed through the valleys, their surfaces cracked and translucent like blue lava. After a few minutes, Okpilak emerged, dwarfing the other glaciers. But the bleached walls of the valley revealed how high Okpilak had once flowed, and made today's ice look like the dregs of a draining bathtub. Its long tail was naked and blue, and dotted with mounds of debris—the results, Matt explained, of ice-cored moraines thawing and releasing avalanches of ground-up rock onto the glacier below.

We began to fly tight, gridded lines over the glacier. What we were doing, said Matt, was taking hundreds of high-resolution photographs tagged with exact GPS points, which he would later use to generate one of his 3D models. The model would capture the topography and color of the valley with near-perfect accuracy. If he wanted to, he could then zoom on his computer to the point where Leffingwell's cairn had once stood, pan to Leffingwell's perspective, and repeat the 1907 photograph, virtually. He could do all of this, he clarified, without setting foot on the ground.

As he explained this, a strange feeling settled over me. It seemed a small miracle that in just a few generations—the snap of a camera shutter in glacier time—humankind had brought one of the most remote reaches of the world to our most exacting virtual scrutiny. Exactly the kind of miracle we'd need in order to register the tectonic shifts rumbling in our midst—from the retreat of glaciers and ice sheets, to abrupt permafrost thaw, to the northward migration of forests.

But beneath this awe for Matt's work, I noticed a feeling of detachment. Just a few weeks before with Roman, the mountains south of here had opened a world to me; now these ones passed beneath me almost like video on a screen.

This feeling no doubt had partly to do with our mode of travel. As the Swiss-French architect Le Corbusier observed in his 1935 book *Aircraft*, flying tends to create mental and emotional distance between the passenger and the world beneath: "I understand and ponder," he wrote. "I do not love." But there was something else going on, too. The sight of the melting glacier tired me somehow. *We all know this story*, said a voice in my head. *We've seen it a hundred times before.*

*   *   *

Back in 2007, Matt manually repeated Leffingwell's photograph from near where the cairn had collapsed, and the resulting photo pair—Leffingwell's, which showed America's largest Arctic glacier, and Matt's, which showed a valley of bleached rubble—became a poster child of climate change, even hanging in the Obama White House. In the years since, images of melting glaciers have flooded our world. Today everybody knows that the glaciers are melting. The story is so familiar that it's become weirdly easy to forget that glaciers themselves are anything but.

NORTH TO THE FUTURE

Glaciers are *wild*. By definition, they are masses of ice so large that they move like rivers, flowing from high ground to low. When you travel on a steep glacier, you can feel it cracking, popping, booming as it churns down the mountain, grinding even the hardest rocks to powder. Some glaciers are so massive that they compress Earth's several-mile-thick crust down into the mantle, causing the ground around them to bulge like a waterbed. Every glacier on Earth predates modern civilization, and there is ice in Antarctica that is older than the human species, and though we may think of glaciers as vulnerable—indeed, we are chipping away at them every year—some will no doubt persist no matter what we do. One day, perhaps after our species has died out, they will creep back over the planet, because ice ages are not merely a thing of the past.

"A man who keeps company with glaciers comes to feel tolerably insignificant by and by," wrote Mark Twain in *A Tramp Abroad*.

> The Alps and the glaciers together are able to take every bit of conceit out of a man and reduce his self-importance to zero if he will only remain within the influence of their sublime presence long enough to give it a fair and reasonable chance to do its work.

But today, for the most part, we don't give glaciers (or much else, for that matter) a fair and reasonable chance to do their work. The very qualities that make glaciers instructive—their ponderous silence, their visceral scale, their temporal weight—are precisely those qualities that digital media flatten. Trying to translate a glacier to a screen is a bit like trying to teach Buddhist meditation techniques over TikTok, or attempting a nuanced discussion over the platform formerly known as Twitter. It doesn't work. The pictures and videos of glaciers keep getting crisper, easier to share, and more "realistic"—but their effect, paradoxically, is to fuel rage and anxiety. The images remind us of political rallying cries and deranged debates; BBC documentaries and doomsday predictions; blowhard contrarians and seething activists. But they do not evoke the glaciers themselves. So the great ice recedes, perhaps even faster from the public mind than from the actual world.

We need these images, of course. They are among the best tools available for communicating about places few people get the chance to visit. But the more readily our technologies allow us to shrink reality down to a screen, the easier it becomes to overlook all the screen fails to convey. This may partially explain why most of the world's major religions historically forbade images of God: Clergy recognized the stubborn human tendency to confuse representations for reality. They knew that to concretize the ineffable in a manmade object would be to collapse its essential *other*ness, to make definite and small something large beyond reckoning.[2] The kernel of wisdom behind these oppressive policies mostly forgotten, we now represent everything—and with such marvelous exactitude that we tend to forget we are even doing it.

The risk is that we will grow more and more content to stay inside and trade the world for images of it. That even when we do get in front of a glacier or the Grand Canyon or Machu Picchu, *In Real Life*, we find ourselves nonplussed, because we think we've already seen it a hundred times before. We look just long enough to snap a few pictures and say we were there, and then turn away.

*   *   *

As we flew back and forth over Okpilak, I caught myself fitting it into the old climate story. It was easy enough to do; the ice had retreated

---

2   The philosopher David Abram explores this thesis in his wonderful book *The Spell of the Sensuous*.

NORTH TO THE FUTURE

more than a mile since Leffingwell's time, and in places had lost more than three hundred feet of thickness. Like all the other Romanzof glaciers, Matt said, Okpilak would likely disintegrate entirely by the end of this century.

But another part of me—my inner Roman, maybe—tried to remember how it had felt to be on the glaciers of Denali. How might Okpilak have seemed to Leffingwell, clambering onto the ice in 1907, a world away from his native Illinois? Or the stray Iñupiaq hunters who no doubt wandered up into this valley a decade, or a century, or a millennium before? Was it too kumbaya to consider how this valley might seem to the wolverines who explored its recesses, or the birds who rode on its thermal winds, or the bears and wolves and sheep who drank from its headwaters?

4

Approaching such questions would require landing on the glacier, I knew. For them to amount to more than fatuous projection, I'd have to get out of the plane and look around. Matt needed to land, too; for all the data he could gather from the air, he still had to travel the glacier on foot to check the stakes he'd planted across its surface. (These show how fast different parts of the glacier are flowing—crucial information for explaining trends in a glacier's volume.) But that wasn't in the cards, not today at least. When we arrived at McCall around noon, having spent an hour mapping Okpilak, we found that Matt's usual landing spot was blanketed in snow. Expecting that it would have melted out by this late in the summer, he'd kept wheels on the plane instead of installing his skis, and now he wasn't sure how to proceed.

"I'll have to figure that one out," he said. "Maybe I can work up the nerve overnight."

So over McCall we flew, continuing north toward the Arctic Ocean.

NORTH TO THE FUTURE

* * *

"Kavik River traffic, Stationair One Delta Charlie," said Matt into the radio, breaking the silence after several brooding minutes. "Twenty miles to the northeast. Maneuvering over the Sadlerochits."

A woman's voice crackled back. "Are you gonna be doing your glacier thing or spending the night?"

"Spending the night."

"Alrighty. The Nolan Suite is available."

We dropped out of the mountains and shot out over a vast expanse of coastal tundra. There were no trees here, and without their scale, it was difficult to tell whether the tussocks and willow shrubs were the size of marmots or mammoths. (This loss of scale is a common experience among visitors to the tundra. The twentieth-century Canadian explorer Vilhjalmur Stefansson once described stalking a bear for an hour, only to realize that it was a marmot; a nineteenth-century German sailor named Johann Miertsching watched hunters approach a polar bear that "rose in the air and flew off": a snowy owl.)

On the western horizon, a cluster of angular, neon orange blobs appeared, hovering over the greenery like a hallucinatory oasis. Gradually the blobs turned three-dimensional, and I saw that they were buildings, skirted by a small islet of gravel. Matt aimed for a gravel landing strip protruding from the north side.

This was Kavik River Camp, Matt explained—one of three refueling stations on the north side of the Brooks Range, a region called the North Slope. The camp was created in the sixties to lodge workers searching for oil, but since 2009, lifelong Alaskan Sue Aikens had leased the huddle of Quonset huts and trailers from the state to host scientists, film crews, and anyone else with the resources and inclination to visit.

The landing was smooth, and as we taxied over to a gas tank, a burly woman emerged from a Quonset hut.

"I'm Sue," she said, opening her arms as I crawled out of the cockpit. "I'm a hugger."

She was a hugger, I soon gathered, when there were people around to hug. As a child, Sue had wanted to be a lighthouse keeper. Now fifty-seven, she had more or less gotten her wish. Visitors generally cleared out by around September, and from then until May, her closest human neighbor was thirty-two miles north, at the Badami oil field. The nearest "big city," Fairbanks (population 32,000), was 334 miles south. "If you lose a finger out here, you might as well stick it in salt and put it on a keychain, 'cuz you ain't gettin' that fucker back on," said Sue's brother-in-law, Rick, a former ranch hand who now helped her care for the property.

Rick had spent much of the previous winter here alone in Sue's stead. The sun didn't rise above the mountains to the south for two months, and temperatures dropped as low as –52°F. The wind blew so hard that it once kept him up for three days straight; he thought his semipermanent shelter might blow down. "I don't mind being alone, but this was different," Rick recalled. By the time a plane arrived to pick him up at the end of his watch, his bags had been packed for days.

Sue starred on *Life Below Zero*, a National Geographic reality TV series that followed several aggressively bootstrapping Alaskans as they did things like set traps and make perimeter checks. On-screen, her every passing comment seemed punctuated with a bullet hole:

"If it hurts, don't think about it."

"You gotta keep moving or you'll freeze right in place."

In person, she spoke with similar gravity:

"How much toilet paper do you use in a year? How much salt? Pepper? What goodies do you want? How you gonna pay your bills? Something happens to this"—she points to a gas stove—"how you gonna get your lumber? Everything is stacked against you. Throw something else on me! What the fuck else you got?"

"They play it up for the show," said one pilot, "but she really knows her stuff."

"She's still coastin' off that bear attack," said Rick.

One fall a few years before the first season of *Life Below Zero*, a young grizzly bear had attacked Sue near camp, puncturing her skull with its teeth and tearing her hips out of their sockets. Satisfied that it had asserted its dominance (according to Sue's interpretation), the bear let her go. She crawled back to camp, cleaned and sewed up her injuries, and tied a belt around her waist to keep her hips in place. Then she grabbed a gun.

Concerned that the bear would return, she ventured back onto the tundra to kill it while she still had her wits about her. It didn't take long to find the animal and shoot it dead, but her hips gave out in the process, and she had to drag herself through a river back to camp. She called state troopers for help, but no one answered. Ten days passed before someone flew over camp, realized something was wrong, landed, and found her.

"I got caught with my panties down once," said Sue, who still had gouges in her skull from the incident, which she let visitors feel. Now she stashed baseball bats around camp. She slept with a .44 pistol.

Her new fame had also brought human dangers. One fan snuck into her trailer and waited for her to return. "You haven't said you love me yet," he told her, brandishing a knife.

"He brought a knife to a gun fight," recalled Sue. She put her .44 to his head.

Sue lived in "the Twinky," a neon yellow Quonset hut with a happy face painted on the side. It used to bunk ten oil men; now Sue shared it with a four-pound, kombucha-drinking poodle named Little Bob.

Bob was a gift from her granddaughter, who lived in the Lower 48. Whenever Sue let Bob outside, she had to watch him so he wouldn't get snatched by a raptor.

"You have no sense of self-preservation at all," cooed Sue, cuddling Bob like an infant.

All in all, Kavik's facilities were less primitive than a *Life Below Zero* viewer might expect. The accommodations were a bit grim—the Nolan Suite turned out to be a windowless box with meat-locker doors, like the rest of them—and the restroom facilities consisted of two plywood outhouses equipped with bullhorns, in case a bear trapped an unlucky guest inside. But Kavik offered hot showers, a laundry machine, and satellite internet—which cost $5,000 per month, according to Sue.

For dinner the night we arrived, we were served fresh salad with our pasta, the ingredients for which had been flown up from Fairbanks. The mess hall refrigerator was stocked with Vita Coco coconut water, LaCroix, clementines, and Chobani yogurt. Sue owned nine freezers, in which she stored at least nine months of food at a time. She had enough pecans to share with the ground squirrels and enough meal leftovers to share with other scavengers of the tundra.

After dinner, as I returned to the Nolan Suite, a red fox dashed by me and slipped into the weeds between two trailers. By the light of the ten p.m. sun, I noticed something oddly familiar in its mouth and took a few quiet steps forward to confirm. My eyes had not deceived me: There, clutched between its feral incisors, was a slice of pepperoni pizza.

5

When I awoke the next morning around six a.m., Matt was already up and checking the flying conditions. Our fears from the day before had come to pass: Smoke had swept north over the mountains in the darkless night, shrouding the camp in acrid haze. The sun was so dim that I had to remind myself not to look directly at it.

"A little smoky out there, huh?" said Rick in the dining hall.

We ate breakfast—coffee, hash browns, and eggs with some kind of meat in them—and then waited for the visibility to improve. Around nine thirty, Matt started packing up the plane. The air was still thick with smoke, and the likelihood of landing on a glacier seemed slim, but he was apparently going to try.

Sue emerged from the Twinky to send us off.

"Go get him, Bob!" she said, sicking the poodle on me. "His Princeton ass won't save him here!"

Once in the plane, we began trundling forward over the gravel and before I knew it—not more than five or six seconds—we were airborne. Matt flew low over the Canning River, which marks the Western boundary of the Arctic Refuge, and from there he turned inland toward the Sadlerochit Mountains, which would guide us up into the Romanzofs. But after a few moments in this direction, Matt laughed.

"Those aren't the Sadlerochits," he said, looking at the smoke-wrapped mountains beyond his window. They dropped out of view as he banked. "I was going up the Ignek Valley." My egg-filled stomach churned. Matt had flown this route about as many times as he'd visited the Romanzofs; if he couldn't even see which valley we were in, then how, exactly, did he plan to land on McCall?

"We're not gonna land on a glacier in anything like this," he added a few seconds later, to my relief. "This is instrument flying."

Rather than fly back to Kavik, however, Matt banked northeast, around the front of the Sadlerochits. As we swooped low over the coastal landscape, I got a view of a place that had thrust Matt into national headlines the summer before, after he'd devoted several months to mapping it.

Before us, as far as the smoke permitted vision, stretched smooth, undulating tundra. The land was webbed with fractal, polygonal cracks, where the earth had contracted in the winter due to plummeting temperatures, then filled with water in the summer, then been wedged further apart in winter by the freezing water—and on and on, over and over again, for millennia. The resulting shape of the land was so empty of scale, so devoid of ocular footholds, that I found my mind swimming, for pockets of time, in abstract forms and ill-defined sensations. Every five to ten miles we passed over a gravel-bottomed stream or river flowing north from the snowy mountains. Some of these were chalky with glacial sediment; others were so clear that I had to look for riffles to know they weren't dry.

This place is known as the coastal plain of the Arctic National Wildlife Refuge, but much of it isn't really a plain. Near the mountains, the rivers flow with enough erosive power to carve deep banks. In winter they trap snow drifts that harbor the highest concentration of maternal polar bear dens in Alaska. The mountain winds offer mosquito relief to caribou, some two hundred thousand of which make one of the longest land migrations on Earth each spring to bear their young here. The mountain glaciers keep the rivers flowing steadily through late summer, nourishing deep pools of fish and skies full of migratory birds from six continents. And the mountain tectonics compress ancient marine fossils into vast, accessible reserves of liquid gold.

It was more than a century ago that Native Alaskans told Leffingwell of hot, flammable liquid steaming out of the ground even in the depths of winter. Leffingwell investigated and confirmed their reports, though he doubted the practicality of drilling in this hinterland. But the calculus changed after 1968, when petroleum prospectors struck the largest North American oil field to date in Prudhoe Bay, fifty miles west of the refuge. The pipeline was built, and suddenly the refuge didn't seem so far away. A decade later, when the federal government doubled the size of the refuge, it also raised the possibility of drilling in the 1.5-million acre coastal plain, authorizing a seismic assessment of its oil potential. The results of those tests indicated that somewhere between 4 and 12 billion barrels of technically recoverable crude oil lay beneath the refuge—less than Prudhoe's 13.7 billion, but still a major windfall. Thus began an all-out war between developers, conservationists, and locals that has raged for decades.

Almost everyone I'd met in Alaska had a strong opinion on the matter. "All the older guys, they lived in the days before there was any money around—and boy, it was just a hardscrabble life," said one resident of Utqiagvik, one of eight communities in the North Slope Borough, which receives about 95 percent of its budget from local taxes on nearby drilling. "Now we can build beautiful schools and have a huge search and rescue helicopter. And if we don't keep development happening, it's all gonna dry up and go back to the old days."

"It is the largest block of pristine land in public hands—in any hands—in America," said Niel Lawrence, Alaska Director of the National Resource Defense Council. "There is no other piece of land that size, that intact, that ecologically functioning, in the country."

Kenji found the hypocrisy of city-dwelling helicopter environmentalists exasperating. "Their life today very much depends on oil," he'd told me that winter. "I am one hundred percent sure they will develop the refuge. So, why not do it soon?"

"It's a fuckin' National Wildlife Refuge!" fumed Roman, when I'd needled him on our trip. "Go drill for oil in the petroleum reserve! It's like being a fat man and then just deciding to rub food all over your body because it feels good."

In 2017, with Prudhoe running dry and Alaska's economy at the bottom of a well, Congress under President Trump opened the coastal plain to oil and gas development. The Department of the Interior drew up plans to conduct thirty-eight thousand miles of seismic exploration to gain a more precise estimate of the value held in the region—and with this in hand, they would auction drilling rights.

Matt claimed to have no dog in the fight. Like Kenji, he had an allergy to hypocrisy that bordered on pathological, and he scoffed at the rhetoric coming from greedy developers and outside conservationists alike. But he'd been doing research in the refuge for fifteen years, and knew that seismic testing was not so benign as the public had been led to believe. He knew how counterintuitively fragile the tundra was—how small disturbances at the surface could trigger chain reactions beneath. Back near Kavik, he'd pointed out the cockpit window to the scars of seismic campaigns from the 1960s, where workers had disturbed the tussock grasses that insulate the frozen soil. The soils had thawed, the resulting puddles had absorbed more sunlight, and the soils had then thawed deeper. Half a century later, gravel roads were now strings of ponds stretching to the horizon. The two-hundred-foot-tall initials "G.S.I."—which Geophysical Service Inc. had carved into a riverbank to guide helicopters—had deformed into a grotesque bog, like a football-field-sized ad to outer space.

The oil industry had, in many respects, vastly improved its environmental standards, but Matt had surveyed its most recent seismic work, and he knew that repeating the process in the refuge would stamp a 200-by-200-meter grid on the tundra that might remain for generations to come. More importantly, it might cause cascading hydrological and ecological changes. So in 2018, Matt cobbled together all of his savings and devoted his entire summer to producing a map of the coastal plain. The idea was to surveil this place: If corporations harmed it in any way, he, and the public, would know about it.

"People are usually better behaved when they know someone is watching them," he told me now. "If a serious company did this with lidar [a more expensive remote sensing technique involving lasers], this would be a million dollar project, easy. Which means that, from the oil company perspective, they didn't have to worry about some bozo making a map without them. From their perspective, they were the only ones who could afford to do that."

In a way, they were right. Mapping the refuge drove Matt to the brink of financial ruin. Not only did he skip prime contract season for his mapping company, which had been paying his bills and funding all of his glacier research, but his direct expenses approached six figures. When I'd met him during the propeller repair the summer before, he'd mentioned that he was "making liberal use of credit cards" and that he'd taken out a loan against his only source of income—his plane.

But his efforts attracted national media attention and reframed public discourse on the refuge. Seismic exploration had been scheduled to begin a few months after Matt made his first map, in the winter of 2018–2019, but *New York Times* coverage of Matt's work showed stakeholders that significant environmental questions had not yet been resolved. By the end of Trump's first term, seismic exploration had still not begun, and upon taking office in 2021, Joe Biden immediately imposed a temporary moratorium on all leasing activities. Environmentalists had won a fragile victory in the war over the refuge—and Matt's research might have made the difference.

"There's no question that, to the extent that public awareness has played a role in that, Matt's work was the essential catalyst," said Lawrence. "There's nobody else out there doing this. It's Matt or nothing."

*   *   *

As we glided over the refuge, the same feeling of unease I'd noticed over Okpilak Glacier was still there. I thought I should be encouraged by Matt's efforts, but instead found that a part of me resented them.

It's not that I doubted the value of the maps; they plainly filled a hole in our understanding of this place. The project seemed deeply admirable, even necessary. But I also knew that had I never come to Alaska—had I instead remained in suburban LA, catching glimpses of "nature" only through YouTube clips and headlines—technologies like this would have shaped my sense of our planet.

They were just tools, which we were free to use or not as we saw fit. And yet the image they collectively rendered would have seemed so marvelously lifelike, so convincing in its objectivity and completeness, that it would have been easy to believe on some level that humans *really had* captured America's greatest wilderness, down to the last tussock and rock and river bend. That Earth was essentially a known quantity, reproducible on a screen. Even now, I had to remind myself of all that Matt's map missed: the frigid darkness that had silenced me in Kenji's winter forest; the Brooks Range wolves, who had howled me out of my head. These forces which had revealed, as nothing else had, the edges of my own life.

Years later, I would encounter a short story, written by Jorge Luis Borges, that evoked the same unease I felt in Matt's plane. Written in 1946, "On Exactitude in Science" told of a fictional empire that achieved such cartographic perfection that it eventually produced a map "whose size was that of the Empire, and which coincided point for point with it." Before anyone realized what was happening, the map had covered over the land, smothering reality itself.

Borges saw that as our representational powers grew—as we developed more powerful technologies and concepts and models to mediate experience—we risked cutting ourselves off from the realm of mountains and rivers, plants and animals, wind and soil and sun. Instead of interacting with these things directly, we would become lost in an endless, abstract conversation with ourselves.

As we circled back toward Kavik, I felt like this was happening to me. I wanted to get out of the plane; I wanted to walk over the tundra and up to Okpilak. I didn't want to do any more scanning or photographing or journaling. To my overwrought twenty-one-year-old brain, a binary choice presented itself: Would I stay in Alaska at the end of the summer, close to the wild and people who could help me see it more clearly? Or would I return home to carry on with my old life?

Over dinner that night, I asked Matt about his early days in Alaska, and how he'd built a life here. He told me his arrival story—how he'd come with friends in spring, and had decided to stay when they'd returned south in fall.

"That must have been scary," I said. "Exciting, but scary."

Matt raised his eyebrows. "It was the thought of going back that was scary."

6

Over the following week, Matt and I made daily attempts to find a way around the smoke to the glacier. Each time, poor visibility turned us back to Kavik. Rick began calling Matt "Tinkerbell." He'd flap his arms and cluck like a chicken when we returned to camp.

"They'll see a mosquito or a cloud and they'll be back," said Sue one morning as we prepared to fly.

She acknowledged, however, that the conditions were unusual. Rick admitted that he had never before seen so much smoke at Kavik. We would later learn that about 1.8 million acres burned in Alaska that month, one of the highest totals for any month in state history.

For the most part, the smoke and mosquitos kept us indoors. The trailers, designed to trap heat, grew hot and stuffy in the afternoons, so we spent much of our time in the dining hall, where the diesel-powered stove was off and two fans were on. The dining hall—half of a caribou-rack-adorned Quonset hut—was lined with plastic folding tables and shelves filled with detective novels, blockbuster movies, and board games. For the most part, people left these alone, opting instead to talk in hushed voices or scroll on their phones. We tended to treat the four-person tables as personal ones.

There were three helicopter pilots staying at Kavik, two of whom were waiting for clients who couldn't get here because of the smoke. The third was working for a four-person team from the US Fish and Wildlife Service, here to take baseline soil, water, sediment, and moss samples before oil and gas development began in the coastal plain. For two days, poor visibility prevented the team from flying even to their nearby research sites. One scientist knitted a sweater, another a baby hat.

"You just totally missed it!" said a third as a pilot entered.

"I missed what?" the pilot replied.

"Absolutely nothing."

The pilot was a blond-bearded millennial named Nick. He'd been at Kavik for about a month now and was "bored as fuck." I remarked to him on the irony of being cooped up indoors in the midst of a wilderness people pay small fortunes to visit.

"Who comes out here for fun?" he replied.

Nick struck up a rivalry with Rick over cribbage, the rules of which seemed up for cuss-filled negotiation. Rick, a baby boomer who saved his dentures for the front country, drew from an ample stockpile of off-color trash talk—and Nick dished it right back. Rick called Nick "Dick," and Nick called Rick a "dirty old man." Occasionally Rick would leave the room to smoke American Spirits; Nick would leave to vape.

At some point I queued up an episode of *Life Below Zero* on the flatscreen TV. Others tuned in. We watched Sue barge into the room we were now sitting in, with a headlight and rifle, to search for a winter bear. When the show referred to Nenana, the home of one of its characters, as "138 miles south of the Arctic Circle," a member of the Fish and Wildlife team snorted.

"Not, 'forty-five minutes from Fairbanks,'" she said. "Not, 'a great place to get a hamburger.'"

Another character, who lived along the Haul Road, declared that "meat is of the utmost importance here in Northern Alaska." "Yeah," muttered Matt, "'cuz it's a three-hour drive to Fred Meyer." Another member of the *Life Below Zero* cast lived just around the corner from Matt in Fairbanks. We turned the TV off.

I left camp that afternoon to walk on the tundra, and after half a mile or so noticed a brown lump up ahead. It wasn't moving, but Sue's grizzly altercation was still fresh in my mind, and I beat a triumphant retreat. Back at camp, I mentioned the encounter to Matt. "Young Ben here saw a bear!" he called to Rick. Jumping onto one of Sue's 4×4s, the three of us motored out to the place.

As we got closer, the lump came back into view, and something shriveled inside of me: The "bear" was actually a small willow shrub—a figment of my hungry imagination. We lit up cigarettes, and Matt and Rick acted like I hadn't made a fool of myself, but I felt like I was back where I had begun, a kid who spent too much time inside.

* * *

For the most part, people retreated via screens to their lives back home or online entertainment. There was a sign on the inside of the dining hall door that said SHUT THE DOOR TO OUTSIDE PLEASE—and in Alaska, *Outside* refers to everywhere that isn't Alaska. But closed though the door remained, the Outside had found its way in.

"When I first started coming up here in oh-ten, it wasn't so bad," said Rick. "Now everyone's like—" He stared wide-eyed at his hand, as if hypnotized.

NORTH TO THE FUTURE

Matt installed a Wi-Fi extender one afternoon, spreading internet connection from the dining room across the whole camp. Suddenly we didn't need to gather in the dining hall to go online.

"Now I can check my phone while I take a shit," observed Nick.

Another afternoon Nick speculated that he had probably walked less than thirty miles over the course of the previous month, and he and another young mustachioed pilot turned to their phones to find the answer. In the process, both seemed to get distracted, and the conversation ended. As they slipped into their phones, the screen on my own phone went black.

I tried holding the power button. I tried charging it. I tried plugging it into a computer. Nothing worked.

I took the phone to Sue to see if she had any ideas. She saw my iPhone 6 and laughed. "A little behind the times, eh?" Sue owned the XS, Apple's then-latest model.

By the time the smoke cleared enough to fly, Matt and I were running low on food and were forced to return to Fairbanks. We never landed on McCall Glacier. And despite Matt's hyper-accurate maps of the region, I felt more lost than before.

7

The day after Matt and I returned to Fairbanks—after he'd landed us safely, driven us to get noodles with his ex-wife and thirteen-year-old son, and dropped me off at my hostel—I had a blow-up with my parents.

I'd never fought with them before: not when the hormones had kicked in during middle school, not when I'd started staying out late drinking in high school, not when I'd told them I wanted to take a year off before college to learn how to be outside. That's because my parents happen to be some of the most decent, even-tempered people one could wish to meet. But over the phone now, they asked about job and fellowship applications. They spoke in tones that said I was the same child I had always been, and that Alaska trips or no, I hadn't changed in their eyes: Soon life would carry on as before. Suddenly I found myself shouting at them—*I don't care about applications, I'm not going to apply to anything*—because I knew I loved them too much not to return home. I shouted at them out of fear that I wasn't strong enough to bring these two worlds together, and that eventually this new, old world might fade from my life like a vivid dream.

After the call, I spent several hours struggling to compose a more considered email to my parents, but the anger and confusion still roiled. Closing my laptop, I changed into shorts and ran out the door.

I ran down the tree-lined streets of northern Fairbanks, up the university's campus, and onto a trail leading into the woods. And as I approached a bend in the trail, I passed a pile of grizzly scat, and something rose suddenly to the surface—a knowledge that seemed to have always been there but had never made itself known. Time telescoped; hot flashes of memory and feeling tore through my body; and I felt, with utter certainty, that a bear was about to appear from the trees and kill me.

In retrospect, it's hard to account for this faulty premonition, because the bear scat was of the same day-or-two-old vintage that populates any Alaskan trail, and I was certainly in less danger than I had been sleeping in Roman's food-stuffed tent, or wrangling Kenji's reindeer, or stepping into Matt's plane. Maybe going home felt like a kind of death. Perhaps after months of exposure to this unforgiving land—months of being broken down and turned inside out and reminded constantly of the passage of time—I had finally reached a kind of breaking point. It was as if, after months of trudging callously into harm's way, all of the apprehension I should have felt before emerged at once, in the form of an imaginary bear. Whatever the reason, waves of memory surged through me, and for the first time in years I began to cry.

I saw my friend Max, whom I had met the first day of college and felt I had known all of my life. I relived his jaunty walk, the warmth in his dark eyes, and the wild joy when we'd decided, one spring afternoon, to blow off schoolwork to see Radiohead in New Orleans.

I felt the presence of my mother and recalled a photograph I had seen of her when she was seven or eight, looking over her shoulder as she ran down a dirt road, laughing back at me through time. How was it that this girl—who had grown and suffered and loved, who had brought me into this world and written books and fought through two rounds of cancer—how was it that she, too, would one day be gone forever?

Memories continued to well up long after I realized that there was probably no hungry bear waiting in the trees. Because whether the end came that afternoon, or decades later, it was coming all the same. The weight of this certainty split me open. For so long, I'd been holding on so tight, scared of losing what had never really been mine to begin with. It felt like I'd been watching myself on a screen, clenching and holding my breath for the final outcome—as though that outcome could somehow save me from the not-enough-ness of my own life. But now the screen had flickered; the frame fell away; and from every direction came pouring in a world that had almost nothing to do with me. It turned out I was just about the least interesting thing going. In rushed sounds and smells and sensations, depth and movement and color, trees and birds and sky. And in that moment, it didn't matter that I wasn't enough, because the world so manifestly was. I wanted to let as much of it in as I possibly could.

After maybe ninety minutes of running and walking and blubbering, my nervous system was shot and my stomach rumbled, so I returned to the hostel to eat a calzone and book flights and carry on with life. But I have not been quite the same since. What I had stumbled upon, on that run, was the edge of the map, and it was far closer than I had imagined.

Life would always be wild, no matter where or how I lived, because death, in the end, could not be photographed away, or modeled into submission, or written out of our story. Death cut through the screen's false promise of separation and control. It was a touchstone more solid than any patch of ground, a trapdoor back into the mystery of being. Once I realized this, I felt ready to return home.

*　　*　　*

Four years after our attempt to reach McCall Glacier, Matt told me he'd made a breakthrough. We'd kept in touch, hopping on the phone now and then to discuss recent findings or his latest funding schemes, like taking money out of his 401(k) to build Airbnb units on his property. ("This year I didn't go out in August because I couldn't afford it," he'd once admitted. "There was just no way.") Now, after more than two decades of research, he'd discovered that McCall Glacier was, counterintuitively, getting *colder*. Its rate of downhill motion was beginning to slow.

The discovery was not inconsistent with prevailing glaciological theory. Ice in a glacier's accumulation zone—which on McCall had disappeared entirely—tends to be relatively warm, because the snow that collects there serves as a heating mechanism.[1] The result is warm, gooey ice that drives glaciers to flow at high speeds—and the speed at which glaciers flow determines, to a significant extent, how quickly they shrink.

So it stood to reason that as multiyear snow stopped accumulating, glaciers would grow colder and slower, and thus their accelerating rate of melting would—all other things being equal—begin to level off.

---

1 In winter, snow insulates the glacier from −40°F temperatures, and in spring, when the top layer of snow begins to melt, the firn (the bottommost snow that's in the process of being compressed into ice) soaks up the meltwater like a sponge, keeping it from running off the glacier. The saturated firn then refreezes the meltwater, in the process releasing a burst of latent heat down into the glacier.

But no one had ever documented this phenomenon with the clarity and precision Matt had on McCall, and the implications were potentially profound. Might the melting of Greenland eventually slow, too?

It is still too early to know. There are many other factors to consider. But part of what struck me about Matt's discovery was his description of the process, which he compared to a "conversation" with the glacier.

"Imagine you've known somebody for twenty years, right?" he said. "And then one day this person shares something with you that is so intimate, that tells you so much about them, that you can't look at them the same way again. Your connection has somehow deepened in a way that you can't ratchet back from. That's in a way what happened with McCall this past year."

This "conversation" hadn't taken place *on* McCall. It took place hundreds of miles away, back in Matt's home office as he pored over decades' worth of data.

His interpretation of that data was grounded in decades of direct observation. He'd spent whole days on the glacier watching ice evaporate straight into the air, a process called sublimation; pockets of snow melt just beneath the surface, leaving thin films of ice above that trapped heat like miniature greenhouses; and bits of rock fall onto the glacier, heat up in the sun, and melt wells in the ice until they rest in shadow. ("Observation drives scientific intuition," he'd once told me. "I don't think anyone should become a modeler until they've spent time in the field.") These experiences were also, I suspected, part of what drove Matt to the ends he'd gone to understand and protect this place. ("It turns out it's not just a hunk of ice, certainly not to me.")

But the oppositions I'd previously imagined—between screens and nature, home and wild—clearly didn't apply. The ice, after all, worked on inhuman temporal and spatial scales. It was only through the lens of long-term records that Matt could begin to grasp the bigger picture.

So he'd charted a synthetic approach. He'd learned how to listen to the glacier *through* the maps.

On the same call, I told Matt the Borges story, in which the empire covered its land with a life-sized map. Then I told him the ending: Later generations abandoned the map. They found it useless, and so let it disintegrate back into the land.

"I don't find that ending very satisfying, actually," I said. "Because we *need* maps, today more than ever. But we need to be able to map the world without killing it off inside of us."

There was a silence on the other end of the line.

"We're not very good, as a society, at listening to the Earth," he began. "We're listening to…God knows what we're listening to. We're listening to a model. We're listening to the TV, to the marketers, to Google. We're not listening to what the Earth is actually saying.

"It's not ignorance that gets in the way of our knowledge," he continued. "It's arrogance. It's thinking that we already know. All the information in the world is meaningless without direct experience. Without ground truth."

PART IV

NORTH TO THE FUTURE

You can sit for a long time with the history of man like a stone in your hand. The stillness, the pure light, encourage it.

—Barry Lopez, *Arctic Dreams*

1

June 4, 2021

Down below the bear is moving. Not charging, but not meandering, either.

Until an hour ago he was dozing on the mountainside, sprawled out as though modeling for a rug. Then he stood, stretched, and began digging for roots. Now he moves patiently, pausing now and then to sniff the air, steadily closing the 150 yards between us. The wind ripples his golden fur the same way it ripples the grass, and I am struck, not for the first time today, by how precisely he embodies these mountains. He brings the entire landscape to a single, brilliant point.

Roman first spotted the bear twenty-seven hours ago, a few hours after our bush pilot deposited us on the bank of a large river called the John, near the center of the Brooks Range. We were beginning the longest expedition any of us have ever taken part in. Julia and Russell are here, and so are two of Roman's other former students, Maddy Zietlow and Toshio Matsuoka. I have never heard of a bear hunting a group of six.

For several miles the bear kept appearing from the trees and then melting back into them, apparently trying to figure out what kind of a thing we are. *Not moose…Not caribou…*After six miles without a sighting and thirteen miles on our feet, we thought it was safe to camp. But this morning, fresh tracks circled our tents. Hours later the grizzly materialized up the valley, cutting us off. We waded across a river, hid silent in a thicket of alders, climbed up this mountain, and—after more than an hour without a trace of him—began collecting data again. *Maybe he was just curious*, we let ourselves hope.

Then we crested the ridgeline; I looked over the far edge, and there he was—sneaking up the back side of the mountain, flanking us again. "*Bear!*" I called, and we scuttled up this peak to claim the high ground. That's when the creature lay down, and the siege began.

Now less than a hundred yards separate us from the approaching bear. Roman picks up a rock; it is time to stand.

The bear is about fifty yards downhill when we release our first volley of stones. They all fall short. On our second attempt, Russell sends a softball-sized rock whizzing over the bear's head. The creature pauses. Roman begins to howl—an awful, ragged sound—and the rest of us join in.

The bear looks up, sniffs the air. Then begins lumbering back down into the valley. But still he moves slowly, with the patience of a creature who has never known fear, has only observed it in others.

"Too bad the rock didn't hit him," says Roman now, watching the bear shrink from view. "I think he's going to come back."

It is eight p.m. The sun is still on our necks and will not set for six weeks. But as we watch the bear disappear into willows, I feel we are back in darkness. Even when he cannot possibly see or smell us, even when we are hidden by trees and downwind of him, the bear has known where we will go. It's like he has one of Matt's maps in his head,

NORTH TO THE FUTURE

has memorized every riverbend and gully and tree in these mountains, and knows just how they will funnel our tired bodies. So it was a relief to stand and hurl rocks and shout. It was a relief to draw the bear's intentions out into the open, even if that intention was to hunt us. Now there is nothing to do but shoulder our packs and continue walking.

Several months ago, when I told my parents that I was returning to the field with Roman, my mom asked, "When?" She had learned about the reckless hunger of June bears—not from me, but from my writing, which was worse. There was no way to hide the risks from her now, so I began joking. "I hear the bears particularly like eating boys whose names start with 'B-E...'" We were talking over the phone, and it took me a few moments to realize that she was sobbing.

I am full of guilt as we walk now. I am also scared. When I first came to Alaska three years ago, my biggest fear was that I might make a fool of myself in front of someone like Roman. The hazards of this land, and the stakes of Roman's work, seemed remote. They don't now, and maybe that's part of why I've come back. But this does nothing for the guilt and little for the fear.

* * *

Two years ago, a few weeks after my first Brooks Range expedition with him ended, Roman made a career-defining discovery. Back in Anchorage, he was scanning satellite imagery of the Western Brooks Range for future routes, and north of the highest mountains in the region, nearly ten miles beyond the known tree line, he noticed dark triangles against the snowy whiteness. Shadows. *Could they be trees?*

A month later, Roman flew into the area and confirmed that the shadows had, in fact, been cast by young white spruce. Which meant that the boreal forest was streaming north at a rate never before observed, into terrain that had been free of forests for a long, long time, perhaps since the last interglacial period more than one hundred thousand years ago. It was a bombshell, but Roman told no one outside of a handful of us. Before going public, he wanted to chart the new extent of the forest, piece together how it had gotten there, and figure out where it might go next.

Somewhat to my surprise, Roman had stayed in touch with me after the 2019 trip. A few weeks after it ended, he began sending science updates, news articles, and even future trip invitations. (Most of these exchanges were friendly, others testy, but I was learning to interpret them all as signs that Roman cared.) While there was never talk of sharing a tent again, I had planned to join him on a shorter tree line survey last summer, until an institutional COVID-19 policy nixed my out-of-state travel plans. No matter, assured Roman: the next trip would be "the best of all."

By the time Roman secured funding and nailed down logistics, I had graduated from college and was reporting in the Himalayas. But for the seven months I'd been abroad, I thought constantly of Alaska and the changes unfolding there. ("I feel like I'm pregnant with something," I said one evening to my then-girlfriend, who was not taken by the analogy.) Those first several trips I had muddled through, just trying to get out of my own head; now I felt ready to actually *see* the place, and to more rigorously tackle the question of how it was changing. And of all of Alaska, the Northwest was said to be changing fastest.

A six-hundred-mile journey, from the heart of the Brooks Range to the Bering Strait. Eleven weeks through one of the wildest, most volatile ecosystems on the planet. *How could I say no?*

* * *

"*Bear*," I warn again, twenty minutes after Russell's throw sent the grizzly slinking down into the valley. He must have beelined back up the mountain as soon as we turned away, because he is now cresting the ridge we descended, some two hundred yards behind us. The others whirl; the creature, seeing us see him, begins to charge.

"If you don't have a gun, get behind me," calls Roman, slinging his hunting rifle off his shoulder. Russell is carrying our other gun, a twelve-gauge shotgun loaded with slugs. Dropping packs and reaching for defenses, the rest of us form a phalanx beside the guns: Julia and I brandishing canisters of bear spray, Maddy and Toshio holding rocks.

"We're going to have to move towards him," says Roman, eyes locked on the bear. "*Now.*"

There is no time to explain, and no need. Our only remaining option is to put the fear into this bear—and, failing that, several bullets. "When you come across a mean grizzly," a Koyukon Native told a twentieth-century anthropologist, "either you kill the bear or it kills you."

Forty yards before us rises a steep hillock we just descended, and as we begin marching toward it, the bear disappears behind its curve. My greasy fingers grip the red canister of capsaicin; my neck tingles in the shifting breeze. Bear spray is our second line of defense, after rocks and before shooting, but only if the wind is favorable. If I spray into a headwind, I might just blind us all.

Thirty yards to the top of the hillock; twenty-five. The wind teases my right cheek. *Click, chh-chhk*: Russell removing the safety on his gun, then pumping a round into the chamber. The noise is slight comfort; a grizzly can fight through several gun wounds, and desperation sometimes heightens its destructive power. Only twenty yards remain now, a distance a grizzly can cover in a little over a second. *Something is going to die.* I think again of my mother. The wind shifts back onto my neck.

As the distance shrinks—fifteen yards, ten—the world tunnels to the feeling of the wind at my back and the pulse of my blood and a narrow field of vision floating above the hill. But that field remains clear. When we reach the crest of the hillock, the bear is still some forty yards off, creeping forward one step at a time. Maybe our disappearance cowed him? Perhaps our aggression—an impertinence he's surely never faced—gave him pause? Whatever the reason, he freezes as we continue marching toward him, then coils into his great bulk and begins fleeing down into the valley. Roman begins to yell, and I let loose:

RRRHHAAAAAAAAAAAAHH
RRHHAAAAAAAAAAAAAAAHHH

Two hours later, we are still looking over our shoulders; my muscles are still trembling.

"If we do go to sleep," says Maddy, "I'm gonna have a pile of rocks by my head."

"I'm gonna have a round in the chamber," replies Roman.

* * *

The bear prowls my dreams but does not return. Close to noon the next day, we emerge bleary-eyed from our tents into dazzling light. The sun fires the mountain snowfields and glints the streams; the air is so dry and dustless that I find myself sucking it in extra deep, as if to make sure it's really there.

After breakfast and coffee, we break camp and set off, working as we walk. Our main focus this summer is white spruce trees, but Roman's scientific agenda has broadened since 2019 to include just about every plant that grows here. In the last few years, researchers have made advances in using satellite imagery to quantify the net "greenness" of the Arctic, a proxy for plant growth, and have found that since 1985, the Arctic has generally become greener north of the boreal forest and browner on its southern edge. The finding suggests a planetary biome shift: shrubs and trees and perhaps entire forests moving poleward. But the trend varies widely from region to region, and the resolution of satellite imagery is too coarse to paint a clear picture.

"What's missing from all the satellite data is which species are causing the greening," Maddy tells me, as we squelch through tussocks overrun with young shrubs. (A recent graduate of APU, she took part in last year's expedition and analyzed the data for her senior project.) "We don't know which species or ground cover they're overgrowing. So now we're filling that gap in the data, and we're learning what to expect in the future. Which species will potentially be overgrown or lost? And which will be competing?"

The matter is not altogether abstract for Maddy. A third-generation Alaskan, her grandparents homesteaded a plot of wooded land north of Anchorage in 1966, and for the next half century hardly thought of fire. But the southern fringes of the boreal forest have begun to dry in many places, and in the last five years her family's cabin was threatened on three separate occasions. In 2016 it would have burned down if not for the heroic efforts of a few firefighters. "It looks like someone put a cup over the house and let everything around it burn," she told me.

Our work this summer takes three forms. As in 2019, we study spruce trees at predetermined field sites and as we encounter them in unexpected places. But we also stop once each day to saw down ten old willow shrubs, collecting cross sections to be analyzed back in town for growth patterns. And NASA has given Roman a grant to gather, constantly, a third kind of data: Each time the plant cover shifts—from willow and birch growing atop tussocks, say, to birch and heather hugging feather moss—three of us (whoever is on duty that day) pull out our phones to independently record the change. Later, Russell and Roman will cross-reference our data with individual satellite pixels to try to determine which plants are responsible for the greening.

The method, which Roman designed and calls "pixel walking," has the auxiliary benefit of keeping our minds engaged with the land. We cannot drift off into reverie, cannot let our eyes settle on the tops of our shoes. But for this very reason, our ankles suffer. It is challenging enough to move quickly through this terrain without constantly scanning the surrounding acres and tapping data into our phones; the combination invites injury. Last year, another of Roman's students strained his Achilles tendon and had to be flown out, and yesterday, as we slogged through a particularly hellish field of tussocks, Toshio twisted his ankle badly. Today he walks with a limp.

But we don't have time to slow down. Unlike the first trip, for which Roman budgeted time early on to whip us into shape, this year we are hitting the ground running and have already lost time to the bear. It is a grueling regime, even for a twenty-three-year-old trail runner; how Roman's trick hip is bearing the strain remains a mystery. It is true that he carries a bit less group gear than we do, and where the creeks are deep and swift enough, he inflates his packraft and floats, taking our heaviest equipment so the rest of us can pixel walk with lighter packs. (This method, which he calls "raft-packing," is a godsend.) Still, I can't explain his confidence that good hydration will see him through.

"I'm wearing out," he acknowledged before the trip. "There's no doubt about it. But if you shoot an old moose, the meat is tough. You shoot a baby moose, and you can cut it with a spoon. As my kids would say, I just got that 'old man strength.'"

He's still got that old man grouch, too, but we get along far better now that we're not sharing a tent. This year that honor goes to Toshio, who emerges long-faced in the mornings. Maddy sleeps with Russell, and Julia shares a tent with me.

She and I are both pleased with this arrangement, I think. We stayed in touch these past two years; she sent a book on California flora and fauna when I was in LA and a beautiful watercolor postcard when

NORTH TO THE FUTURE

I was in Asia. Now we spend our dinners debriefing each day, an activity in which we are unequal participants. Mostly I listen greedily as she unleashes a torrent of observations and impressions.

Still, I am not quite so lost as two years ago. Since then, I've begun taking time each day to sit still, close my eyes, and focus on breathing until my mind begins to quiet and the world filters in. Most days I fail, spending the whole time having imaginary conversations or thinking up to-do lists—but even so, this practice seems to have held open a space for the bears, the wolves, and the glaciers. And now, back in the field, I have more or less picked up where I left off.

*   *   *

We climb west into cloud-raked highlands, crossing and recrossing the Continental Divide. Graupel strafes a pair of swans bedded on a saucer lake; thunder rolls as we pass the fresh pink skulls of two caribou, likely ambushed by wolves. On a pass, as we rest in a thin ray of warmth, the valley before us suddenly pulses; a mass of air sweeps up and over us, whistling through our jackets, blasting bits of debris into our eyes. Moments later the wind settles, and Roman begins down into the valley.

Toshio's ankle is getting worse. This is worrying news; we have only two and a half days' worth of food left to reach our first resupply cache on the Alatna River, and our planned route is another fifty miles. So Roman decides we should take a shortcut, down a steep river canyon called the Kutuk, which we expect to offer relatively smooth travel. Roman walked its lower portion in 2012, and in 1976, a well-known botanist named David Cooper—who plans to meet us on the Alatna several days from now—described the Kutuk as a gentle place, marveling at "the timelessness of the life in this valley, the timelessness of the river…"

But if time ever left this valley, it has since returned. Just below the pass, foaming out of a gorge across the river, is a creek the color of rust.

The willows around it look withered, like a dog's been pissing on them for years, and the rocks are stained a dull ochre. Where it meets the colorless Kutuk, the creek turns the river a runny mustard.

"Mountain must have blown out somewhere up there," says Roman, scanning the peaks. He means that permafrost thawed and collapsed, unearthing old stores of minerals. Once exposed to the elements, these deposits are no different from mine tailings: many of them oxidize and release acid and heavy metals into the water table, poisoning the rivers and lakes below.

These "rusting rivers," as Roman calls them, are perhaps as old as these mountains. Several Brooks Range waterways bear the name Ivisak, Ivishak, Ivisaruk—all variations on the Iñupiaq word for iron oxide, traditionally used as a decorative red dye. But their number is bound to increase as frozen landscapes come apart. Along last year's route, red rivers were so numerous that for the first time in his more than four decades in the Brooks Range, Roman struggled to find drinking water. A river called Clear Creek curdled his powdered milk and left him wondering what might be happening to the fish and other animals in the watershed below.

"I never used to see these, and now they're everywhere," he tells us, eyes glittering.

* * *

We camp on the bank of the Kutuk, drawing water from a glimmer of snowmelt tumbling out of the mountains. In the morning, fresh grizzly tracks circle our tents. Hopefully a different bear. Roman inflates his packraft and shoves off into yellow rapids, while the rest of us set out on foot to gather data.

The head of this river is flanked by great sheets of aufeis, on which we walk for as long as we can, savoring the solid footing and pausing now and then to admire the slabs of melting ice. Soon the floodplain narrows, shunting us up into the valley's brushy midsection and forcing us to bash for several miles through a veil of leaves. Where the thicket ends we come to a halt.

"Holy shit," mutters Russell.

Across a broad grass clearing, less than a mile away, a dark mass pours out of the mountains, blocking our path down the valley. In form and scale it resembles a glacier, but it is dark, almost black.

Russell takes off across the clearing. The rest of us string out behind him. Soon the wall dwarfs Russell's burly six-one frame: It must be at least forty feet tall, and is composed of rocks of every size, from flinty chips to body-sized boulders, all locked in placed by some kind of dark slurry. I expect this to be cakey underfoot, but when we reach it, I find that the substance has hardened into a kind of cement.

We climb over the wall's edge and into a sea of rubble, stretching several miles up into the mountains, and at least several football fields across. Spires of thawing earth slowly disintegrate before our eyes, releasing flurries of rock that settle around them in cone-shaped mounds. Between many of these run deep gullies, held open by boulders and ice chunks. I climb down into the shadows of one and wet my foot in a dank puddle; a disinterred bone—I cannot begin to guess how old—protrudes from a wall.

For perhaps ten minutes we wander the labyrinth—peering, touching, sniffing, drifting apart. When I crest a mound, it takes several moments to spot the others, their colorful shirts bobbing in and out of view like sails on a heaving sea. Soon the others notice the distance growing between us, and we motion to reconvene.

"It looks like somebody took a chunk of another planet and threw it down this valley," calls Russell as we draw closer.

"Smells like fresh rock," adds Julia. "And those spires over there would have rounded off pretty quickly. Looks like some of this happened since last summer."

Landslides like this one, called retrogressive thaw slumps, are triggered by thawing permafrost. No one knows exactly how common they are because they mostly go undocumented, but one intensive study found that thaw slumps increased sixtyfold on Canada's Banks Island between 1984 and 2015, with most of that increase coming in the last decade. It would take a fleet of Matt Nolans, mapping the entire Arctic year after year, to produce an accurate region-wide estimate, and even then we would not know how much land is vulnerable to slumping in the future, or how much carbon it might release into the atmosphere. For this reason, abrupt permafrost thaw features like thermokarsts and thaw slumps are still not represented in most climate models, though some experts believe they might soon emit more carbon each year than countries like Brazil and Russia.

We document the thing and keep moving, but over the next few hours we scramble over several similar features, some of which appear alarmingly fresh. I wonder aloud what we could possibly do if part of the mountain gave way above us; Julia nods and tells me that last year they found a fresh moose carcass stuck in mud from a recent slump.

* * *

In the evening, Julia picks up another strange scent. A few minutes later, entering a broad clearing, Maddy spots a lump of brown fur several hundred yards away. We slip wordlessly back into the trees. The tingling on my neck tells me it's a bear, but Julia is less sure. "It didn't smell like one," she says. "Maybe it rolled in a kill."

When we reach our rendezvous point with Roman, Julia smells smoke. I notice human tracks leading off into the woods, and a couple minutes later we find Roman warming himself in a glen. All around him, clinging to shrubs, are wisps of long golden fur.

"I've seen muskoxen in weird places, and this is the weirdest," says Roman. Julia grabs a clump of the muskox fur, sniffs it, and, recognizing the strange scent from earlier, breaks into a bewildered grin. (There are, in all of Northwest Alaska, only a few hundred muskoxen, and they are virtually always found in herds on the open tundra.) We laugh deliriously, intoxicated by the strangeness of this valley. But when we mention the crumbling mountains, Roman's face hardens.

"This whole valley is coming apart," he says. "Sometimes I worry that something's going to fall on you guys out here. Because now it just happens, and who knows when."

In the tent, after gathering water from still beds of snowmelt and cooking dinner, I find my mind unsettled. "It gives me comfort," wrote David Cooper of this valley forty-five years ago. "It seems as if I have been here before…" But when I close my eyes, I think of churning hillsides and tumbling boulders; waves of liquefied earth and suffocating animals. I think mostly of the water. It was the water that filled my imagination all those years ago when I first read John McPhee's *Coming into the Country*: descriptions of swift, clear rivers teeming with wild fish. "Looking over the side of the canoe is like staring down into a sky full of zeppelins," he wrote of the Salmon River, just a few watersheds west of here. (Eighteen months from now, I will receive a package from McPhee containing an article coauthored by Roman and several colleagues: The Salmon River had, sometime between 2018 and 2019, turned a metallic orange. "This came as a blow to the solar plexus," McPhee scrawled across the printout.)

To McPhee in 1976, Arctic Alaska had still seemed a land beyond human history, a system of natural forces held in primordial balance.

"What impresses someone most of all about the Arctic world are its cycles," he wrote. "Meteorological cycles, biological cycles...Cycles unaffected by people. The wilderness operating in its own way." But tonight, in this valley at least, this no longer seems quite true. The land is grinding into itself, bone against bone.

* * *

In the morning Toshio's ankle is worse. No one says it, but we all must wonder whether his campaign is coming to a premature end. With the food all but gone and eighteen miles remaining between us and the cache, Toshio joins Roman in a second packraft, while the rest of us hurry overland, pausing only to collect puffs of muskox wool that string from branches. Softer than cashmere, warmer than sheep's down, this material, known as *qiviut*, is perhaps the finest wool in the world, and so scarce that it is worth more than its weight in silver.

After nearly a week beyond the tree line, the first trees appeared late yesterday afternoon; now white and black spruce, balsam poplar, and paper birch choke the valley and tunnel our vision. There are no caribou trails here. The only paths through are in the footprints of bears, who mold well-worn pads sometimes used for many generations. I am not wild about following bears, but their tracks offer the only relief from bushwhacking, and I trust Julia and Russell to judge—by the hardening of sap on broken branches, the freshness of prints, the pungency of scents—that no bear has passed through here today. I have come to trust Julia's nose almost as much as my own eyes.

Afternoon comes still and balmy. My ears were already itching from sunburn; now the skin has started bubbling. The mosquitos arrive, too. Russell slaps six on his leg, and reminds us—perhaps mostly himself—that this is nothing. "We could be slapping a hundred next month," he says. The upper Noatak River, where we are headed, is known as the buggiest stretch of Northern Alaska.

The valley constricts, and we cross and recross the icy river, linking arms against the current. A mosquito buzzes into my eye and I stumble over a submerged boulder, nearly dragging Julia down with me. A sharp pain cuts through the numbness in my left Achilles. *Not good.* I take three tabs of ibuprofen on the far bank, but the pain grows steadily worse over the next several miles. By the time we emerge out of the tight chute of the Kutuk, into the broad forested valley of the Alatna River, whatever presence of mind I've celebrated this past week has left me. I abandon my pixel-walking duties, following Russell and Julia all but blindly.

Twelve hours after we left them, we meet Roman and Toshio sitting by the packrafts on a mud bar. Roman ferries us one at a time across the Alatna to our cache. When it is my turn to climb into a raft, I slip and plunge to my neck in dark, icy water. I am in too much pain to be embarrassed.

2

June 10: Crunching leaves, unfamiliar voices. I stir from my sleeping bag and unzip our tent just as a wall of shrubs by our camp parts, revealing a cannonball of a man with sun-faded blue eyes and a bloody slash across one cheek. Close behind him follows a willowy young woman, face flushed from the cool morning wind.

Roman greets them warmly and introduces us. They are the explorer Forrest McCarthy and the painter Klara Maisch, two of four guests Roman had flown to a nearby airstrip yesterday to join the next leg of our journey, into the Arrigetch Peaks. Iñupiaq for "fingers of the outstretched hand," the Arrigetch are the visible tips of a granite pluton that extrudes up through the softer limestone and shale bedrock of the surrounding mountains, roughly cleaving the Brooks Range in two. It was here that David Cooper documented, in his first ever scientific publication, the highest known trees in the North American Arctic—and four warm, wet decades later, Roman doubts they still are.

The peaks are obscured now by clouds but lie only a dozen miles to the southwest, up a side valley of the Alatna called Arrigetch Creek.

NORTH TO THE FUTURE

Today we plan to hike eight miles up the creek to where Roman's guests have established a base camp, and from there to stage day trips through the mountains. Toshio is in bad shape this morning; he decides to rest. My Achilles feels like a block of wood, but if half of what I have heard and read about the Arrigetch is true, I can't bear to miss them. I down six hundred milligrams of ibuprofen, gather food from the resupply barrels, and pack my bag.

For the first half mile out of camp we wade through fetid sloughs and bushwhack through boggy stands of black spruce and alder. (It was one of these switchlike alder boughs, says Forrest, that unpinned from another branch and slashed his face yesterday.) Then the climb begins, up through lush evergreen forests that remind Julia of coastal Washington. Moss-covered logs lie across clear rock streams; mushrooms drip from the trunks of ancient spruce. Up slopes of blueberry and lingonberry, the forest thins, rolling over the moraines of prehistoric glaciers, opening onto fields of lichen-covered talus. Ribbons of water tumble out of the mist.

In late afternoon, just as the anti-inflammatories wear off and my leg pain grows concerning, we emerge from the last fingers of the forest. Here Roman's third guest, the renowned Alaska painter Bill Brody, is waiting for us, and says that the fourth, David Cooper, has wandered off into the mountains. After a short break, Roman, Russell, and Maddy shoulder their packs to return to the Alatna, so they can ferry up another load of supplies in the morning. Julia and I will remain here for the night so I can nurse my Achilles.

We pitch our tent on an island in the creek, and I sit down to ice my leg in the flow of fresh meltwater. Up above the mountains are still veiled by clouds, but now and then gauzy holes open, revealing floating walls and airy fins. For several centuries at least, these mountains have inspired religious feelings. According to the Nunamiut, a nation of inland Inupiat who once hunted caribou throughout the central Brooks Range, the peaks were formed when their creator, a giant called Aiyagomahala, placed his great mittens in the snow. "You will remember me," he said, and then disappeared forever.

The first Western description came from a nineteenth-century naval explorer, who wrote of "lofty minarets" and "cathedral spires." David Cooper asked: "What can I feel, what should I feel among the finest mountains in the world?" He was twenty-four at the time, in the midst of his first, month-long trip to the Brooks Range, during which he fished and foraged to supplement his spare rations. "My mind soars with the mountains," he wrote. "My heart finds a place for this sight and tucks it safely away forever."

Finished with icing, I lean over to wash my face, and as I rise, dripping, an unfamiliar figure arrives in camp, old and stiff. I dress and cross the creek to him.

"You must be David," I say, reaching the bank.

In old photos, his eyes are flinty blue, and he sports a shock of red hair that wraps down to a lumberjack beard. Now seventy, his hair is snowy and trimmed, his sunken eyes soft. But they radiate wonder now as he looks up at me, as though he is coming out of a trance.

"It feels like I never left," he says.

* * *

Early the next morning I meet David and Forrest by the creek. Night reaches into this deep valley even in summer, and a few hours ago, when I lurched groggily out of my tent to pee, the sun touched only the tips of the peaks, bathing the stone in rose glow. Now the sun is up and the clouds have returned, but not so thickly as to foil our task: to repeat a collection of David's old photographs, in order to see how the landscape has changed over the last forty years.

NORTH TO THE FUTURE

Forrest—an old packrafting friend Roman enlisted to help with our overflowing research agenda—leads us straight up the side of the valley. He walks with David's photographs in hand, pausing now and then to reference landmarks: a trio of old spruce, a fork in the creek, a notch in the ridgeline. David does not remember where, exactly, he took these pictures, but each stretch of tundra unlocks memories. "This here is *Zigadenus elegans*," he says, pointing to a short, unassuming stalk. "Deathcamas. That'll kill ya." A few steps later: "This is *Valeriana capitata*. You can eat those seeds and cook the root, if you're real hungry. And there's more *Hedysarum alpinum*. Eskimo potato. Man, I ate thousands of those things."

Between 1978 and 1980, David spent a cumulative eleven months in these mountains, cataloguing more than two hundred plant species in order to address one of the great mysteries of plant ecology: How did boreal plants survive the last ice age, when most of North America lay under sheets of ice that reached down into Indiana and New Jersey? He found that the flora of these mountains overlapped far more with mainland Alaska than the rest of the Arctic or North America, a clue that Alaskan ecosystems evolved somewhat independently. Paradoxically, most of Alaska was ice-free in those days; the region was too dry for large sheets to accumulate anywhere but the coasts and mountains. These hardy plants, David reasoned, must have been among the species that clung to life in the ice-free interior, spreading north when the climate began to warm some ten thousand years ago.

As we climb, he reads the land aloud. "Most people think this green stuff is all just a bunch of weeds," he says. "But they live here all the time, and if you know your plants, you can learn how the land was put together." He fingers flowers and pauses to identify birds by their calls. He points out where the bedrock turns from granite to limestone and pokes his hiking pole into abandoned squirrel burrows.

"There used to be millions of small mammals up here. You'd hear 'em constantly. Now listen: silent. The pilot mentioned some big snow years, so they probably either starved to death or suffocated in their holes. That's a big change right there."

I ask if the land has changed in other ways.

"More striking than anywhere else I've seen," he replies. "Trees, trees, trees. They're all over the goddamn place. When I was here there were a handful. Now there are thousands. And the big trees are not really the whole story. The seedlings…there must be millions in these valleys."

After an hour or so of climbing, we appear to be nearing the spot from which David took his first photograph, a downward panorama of the valley. It is difficult to triangulate the precise point, because in the intervening years trees have popped up; the slopes have grown brushy; even the river's course has drifted slightly.

The changes are large for four decades, but not for the long timescales David's thesis explored. Forty thousand years ago, a glacier filled this valley and North America was human-free. One hundred million years before that, the molten rock that formed these peaks was probably still rising and cooling beneath Earth's surface. Alaska would have been a fraction of its present size—a raft of terranes still catching land out of the Pacific, like logs in a dam.

For a few moments we look back and forth between the photo and the valley, trying to orient ourselves in the flux.

*   *   *

In the evening, Roman, Russell, Maddy, and Toshio arrive in camp, and all ten of us—ranging in age from Toshio (twenty-two) to Bill Brody (seventy-eight)—gather beside the creek for dinner. It is the first time David has seen Roman in person since saving his life here in 1979, and as we boil water, Roman tells the story.

Roman was eighteen then, and had just finished a month-long

climbing expedition in the Arrigetch with three friends. As they were breaking camp to catch a bush flight back to civilization, an aggressive bear approached, and Roman shot it. The creature slipped off—and, nervous about hiking out with a wounded bear around, Roman followed the blood trail until he found the animal, and shot her dead.

His friends returned to Fairbanks, but Roman decided to stay and make use of the bear meat. "I thought I was going to live off the land," he explained to us now. "Spend the winter in the Brooks Range. Build a cabin with my ice ax. I was completely delusional." He cached the carcass in a snowbank, but a wolverine dug it out and—presumably to mark its territory—pooped all over the meat, and Roman's plans, too. Down to his last few scraps of food, Roman began wandering around in the rain "like a lost dog."

In the next valley south, he happened upon a tent and found David inside. David shamed him for his foolishness but also warmed him up, shared food, and told him about a group of geologists forty-five miles away who were flying out in a few days. Roman hiked to them and hitched a ride to safety.

Above us now, the Arrigetch are out in all their swooping glory, a glittering labyrinth of stone. Some peaks are faintly dizzying in their vertical lift; others convey the molten weight of anvils. It is funny to think of a young Roman in this place, scared and no less ignorant than I am. In a way it gives me hope. As we finish dinner, he points the mountains out to us—the ones he once climbed, the ones that got away.

"Now I'm just happy to look at them," he says.

* * *

June 12: I wake this morning to a dull throbbing in my Achilles. Toshio has decided to fly out; I worry that I am next. While the others head into the mountains to gather data and retake David's photographs, I remain at camp—stretching, icing, and digging my fingers into the surrounding muscles and tendons until my eyes water. I have a week to recover, or my journey will end.

The plein air painters Bill and Klara work near camp, and from time to time I join them, trying to figure out why Roman has scraped together grants to bring them here, though they make no material contribution to his research.

The older and better-known of the two is Bill Brody. A shorter man with a sun-baked complexion and wild, wiry hair, he has cumulatively spent years of his life in the remote Alaskan wilderness, often alone, to paint. When someone asked last night why he works out in the elements, rather than from pictures, he replied: "Would you rather kiss your honey, or look at a photograph?"

A math major at Harvey Mudd, Bill worked for years building virtual reality software, an occupation he left in 2006 to paint full time.

"I quit doing that so I could do more of this," he tells me now, standing on the bank of the creek, sketching the bones of the Arrigetch on a two-by-five-foot canvas. "I like technology. Humans make good stuff. But nature makes it better."

"Why paint it?" I ask. He takes a moment to consider.

"Being in this place changes what you're like," he begins. "I'm not trying to paint something that *looks* like this place. I'm trying to let the viewer be here a little bit themselves. If there's a definition of successful art, it's that you come away more capable of experiencing the universe. You see the world afresh."

Later I limp over to Klara, who paints on a slope above camp. While Bill's brushstrokes seem to wriggle and writhe on the canvas, Klara's flow like long-exposure photographs. Bill prefers to "see new country," but Klara returns to the same places again and again.

NORTH TO THE FUTURE

"I try to watch change over time," she tells me. "Landscapes have different timescales and rhythms. Granite has its own language. Glaciers have their own language. So do the rivers, or the trees, or these adorable little birds that are popping out everywhere!

"Sometimes I feel like our timescales are too small to really get at this stuff. But being exposed to this constant change, this complexity, this interconnectedness, it bends us in a different way. I want to show that, to show the world in motion. I think that's what I'm trying to solve with paint."

Late afternoon, as I sit icing in the river, I reconsider my own attempts to describe this place. So often I have used language—like photos and scientific concepts—as a mirror. A way of shrinking experience down to something I can control. I wonder now how it might serve a different purpose.

Yesterday, Julia pointed out a bird to me: a small, olive-colored puff called the ruby-crowned kinglet. The anthropologist Richard Nelson saw it often while living with the Koyukon in the 1970s, and when he repeated the Native words for its call—*Chee Chee Chee Chee tahoodzee tahoodzee tahoodzee*—to a nonnative birder, the latter immediately identified their maker. The Koyukon were such keen observers of this land, their language so finely attuned to its lilting rhythms, that their words evoked the birdsong itself.

How might it be, to pay that kind of attention?

* * *

The next morning, Roman and David head into the mountains together. Already Roman has found several trees far above the highest David recorded four decades ago, and when the pair return in the afternoon they have worked each other into a lather.

"We've found spruce seedlings in every habitat!" says Roman.

"Windblown acid soils," adds David, "dry hilltops, calcareous peatlands..."

"...even up in this marshy area seems to work," says Roman. "And then popping out of a snowbank, with lichen and *Cassiope* all over. You know, it's like they can grow—"

"—anywhere," finishes David.

"Anywhere the seed lands," continues Roman, "it seems like they're going to grow!"

Since 2019, Roman and Paddy have analyzed our data and found that the growth of Brooks Range spruce trees is correlated more closely with variations in nutrients than in water. But each is likely critical, and one variable, which the Arrigetch has in abundance, indirectly provides both: snow.

Snow acts like a blanket in winter, keeping the ground relatively warm and allowing microbes to turn organic matter into nutrients. It also protects seedlings from strong winds and provides a water source in spring and summer. And as year-round sea ice retreats from the Chukchi Sea on Alaska's northwest coast, the North Pacific storm track has moved poleward, radically increasing the amount of snowfall throughout Northwest Alaska.

"The east side [of the Brooks Range] has plenty of warm air, but nothing's happening," Roman continues. "Here there's been two decades of warm air *plus* snow, and now you have a baby boom. And that's the sign of exponential growth."

"That's right," says David. "And the best indicator that there's a lot more snow in these mountains is the absence of these ground squirrels. When you come back when you're my age, this will be a forest. All the way up to the goddamn peaks!"

There is a strange look in Roman's eye, like he knows we've seen nothing yet.

NORTH TO THE FUTURE

* * *

June 14: This afternoon we say goodbye to Toshio. After a round of hugs, he and Roman begin down to the Alatna, from where a pilot will fly him to Kotzebue. We are sorry to lose his sense of humor, and reluctant to shoulder his equipment. The coming miles will be harder without him.

My own fate remains undecided. Watching me limp around camp, David shakes his head. "On a normal trip I think you'd be fine," he tells me. "But the way you guys are traveling is like the frickin' Marines."

David has himself suffered Achilles injuries in the past, and hesitantly agrees to perform a treatment he once found helpful. Turning his gaze to the ground, he walks a few paces, bends down, and picks a white spruce limb from the litter. Then he flicks open a pocketknife and, with fast, fluid strokes, begins whittling the wood down to a compact, beveled bar—thin as a finger on one side, thick as three on the other.

He tells me to lie facedown on the ground and flex my foot. Kneeling behind, he places the thin side of the bar across the taut line of my Achilles. Then he puts his body weight into the tendon and begins raking up…down…I feel my face and neck flex. Within a few strokes he stops: My skin is flaking off under the coarse wood, he says. So I grab my hiking pole and ask him to continue with that.

The pain is bad, but when I stand, the tendon is looser. I return to the creek to ice before dinner.

* * *

June 17: Summer has arrived. Across the alpine slopes, the air vibrates with bees and birds, and the alders, naked just days ago, are flush with leaves.

"It's freakin' hot," says Forrest over lunch, as we sit around in our underwear, sweating. David measures the temperature in the shade at 80°F, just shy of the 82°F maximum he recorded across his three summers here.

My Achilles has improved, but it still hurts just walking around camp. Roman makes a proposal: The next leg of our journey—a four-day trek over these mountains to the headwaters of the Noatak River—will be brutal on ankles, but from there we will inflate our packrafts and float for five days. What if I skip this four-day hike and instead fly out to Coldfoot with David, Klara, and Bill? Then I can hitch a ride into the Noatak when the pilot deposits our next food cache in three days' time. It'll give me nine more days of rest before the climb into the most rugged stretch of the Western Brooks Range, the Schwatka Mountains. "It'll probably cost me a few thousand dollars," says Roman, "but it'll be worth it to have you along." Thrilled, grateful, I agree on the spot.

Tomorrow we will all head back down the valley, and tonight after dinner, when the rest of us head to sleep, David stays up. I wonder if he will ever return to these mountains, which he said yesterday taught him "his place in the world." I watch him observing a brood of Lapland longspur chicks in a spruce tree. This week we have seen nesting songbirds in quantities he does not recall observing here before. The climate may be pushing populations northward, and the disappearance of would-be predators like squirrels and foxes—caused, in all likelihood, by the increased snowfall—has made these mountains an avian sanctuary.

"Pretty soon there'll be pigeons up here," he grumbled yesterday, but he seemed happy.

Now he watches the chicks peering over the edge of the nest, working up the nerve to attempt their first flight.

3

June 22: Today I return to the field after three nights holed up in a trailer inn at Coldfoot. The first night, standing hesitantly next to a hallway washing machine, I could not bear to dirty the one clean towel provided me; so, stripping naked, I threw my clothes in the machine and dashed back to my room, slipping behind the door just as footsteps sounded around the corner. Outside, cold rain had begun to fall and would continue for several days. *If I'm going to miss a part of the trip*, I thought in the shower, watching dirt suds drain around my feet, *I'm not mad it's this one.* The steaming water, fresh coffee, and cheesy omelets are not things about which I plan to inform the others.

But three days in a small trailer is three days in a small trailer, and whether my Achilles is healed or not, I won't stay. I meet our pilot, a spritely, bearded man named Dirk Nickisch, at the Coldfoot airstrip, and we take off west in his single-engine bush plane, a blue Beaver. Having flown here since 1994, Dirk is one of the most in-demand pilots in Northern Alaska, and after three days of rain, he's rushing to clear a long backlog of flights.

"Things are changing so rapidly, it's just kinda mind boggling," he explains now through his headset. "Ever since the ice pulled out of the Chukchi Sea and the Bering Sea, our storm systems are bigger. We have four to five times as many no-fly days. You used to know what different weather systems would do. Now you're like, 'Hell, I have no idea.' It's almost like having to relearn how to fly out here."

About an hour into the flight, we skirt the spires of the Arrigetch; a few miles farther west looms the black-and-white pyramid of Igikpak (8,276 feet), the highest peak in the western Brooks Range. On its far side we descend into a lush valley, threaded by a translucent teal line.

"The Noatak?" I ask. Dirk nods.

The Noatak River stretches 425 miles, west to the Chukchi Sea, and drains the largest entirely undeveloped watershed in America. The lower two-thirds of the river lie within the 6-million-acre Noatak National Preserve. Its headwaters are part of Gates of the Arctic National Park, which, at roughly 8.5 million acres, is the second-largest park in the US, and also the quietest, receiving only a few thousand visitors each year. The entire watershed is thought to be empty of trees until it bends south near the coast—but Roman knows otherwise. For the next five weeks, we will weave in and out of the basin, charting the new forest.

The landing is quick: One moment it seems we are about to crash into the river thirty feet below; then a narrow gravel island appears around a bend, and Dirk aims for it. A touch, a small bounce, and the plane skids to a halt. Quickly we unload the food barrels, a shotgun, and my pack onto the gravel, and Dirk paces out a takeoff zone across the small island. "Some weather coming later this week," he warns on the way back, climbing into the cockpit. Then he motors away—his right wheel splashing in the river as he turns the plane 180 degrees, back toward me—and takes off almost directly over my head.

Then, for the first time since the winter at Kenji's, I am truly alone.

I leave my pack and the gun leaning against the barrels and set off to search the island for tracks. Finding none, I wade to the next island, plashing upstream through crystalline waters. Under a warm sun, the current sends white and blue flashes into my eyes; smooth stones warble and vibrate. Over sand-bottomed channels, the water takes on a milky blue glow that I have seen, outside of the Brooks Range, only in tropical lagoons of the Indian Ocean.

A family of wolves has left crisp tracks on the next island; they have been here since the rains stopped. Nearby, half a dozen Arctic grayling swirl in a deep eddy. A passing comment from David Cooper rings in my head: "These rivers are full o' dumb, hungry fish." I turned vegetarian after my last trip to Alaska, but cannot resist the prospect of meat so fresh.

As I circle back to my pack, a horsefly lands to bite my exposed stomach, and I stun it with my hat. Holding it buzzing between two fingers, I impale the fly on a blunt, barbless hook that Klara fashioned for me out of the baling wire she uses to rig her canvases. I tie the hook to a fifteen-foot thread of fishing line I packed on a whim and fix a twig six inches above the hook to serve as a bobber.

Sneaking back to the eddy, I toss the fly—still fizzing—into the gently swirling water. The grayling circle for a time; then one strikes. I rip at the line, and there is a moment of resistance, but the hook springs free, the fly gone. I repeat this process with several more flies, to no avail. Defeated but happy, mind swimming with visions of clear water and fat, sleek fish, I return to the barrels, gathering willow boughs and wisps of dry grass along the way. Back at camp I start a fire.

After dinner, a wind blows out of the mountains; the fire saws. I sit for several hours, placing sticks on the fire, watching the mountains and the way the colors of the land change as the sun swings. Listening to the soft current and feeling the tingle of dried river water on my skin. Several wolves yelp in the distance.

When the pile of sticks is gone and the fire is down to embers, I dash them with water and pitch my tent, in case rain comes. Then I lay my sleeping bag under the evening sun and let the sound of the river lull me to sleep.

* * *

June 23: This afternoon the others arrive. They seem excited to see me, but more excited to get into the food barrels. Our typical drill is to eat as much as we physically can at the cache, and then travel light to the next one, which in this case is ten or eleven days away. Once we have pawed through our own rations—like last trip, they are an orgy of sucrose and dextrose, -ines and -ites—we begin to eye one another's piles like polite dogs.

Soon the haggling begins. Maddy wants Fritos, and Forrest is sick of his fruit and nut chocolate. Roman—fanning out his Cadbury bars like a stack of bills—will trade only for barbecue chips. "I need all of these," he finally says, swiping the remaining bars into his "to bring" pile. Having thrown all two hundred pounds of my rations together in my first two days in Anchorage, my own pile is far and away the least desirable. Cutting meat from my diet did not help. With my baggies of textured vegetable protein, my freeze-dried veggies and vegan gravy packets, I am Dunce, the boy whose mom packed carrots.

While the others continue swapping and munching, I turn my back on the sorry affair to search again for grayling; perhaps a few bites of fresh fish will raise my stock. Julia joins. Seeing my wretched tackle, she ties my line to her hiking pole and removes the bobber to make our rig more responsive. We sneak back to the same eddy I tried yesterday.

For a time we take turns casting, managing only to feed horseflies to the cursed fish. Finally I tug at just the right time, and a flashing orb breaks the water, slingshotting onto the tundra. The blunt lure disengages in mid-air, but before the fish can flop back to safety, Julia dispatches it with a stone. It is eleven inches of shimmering iridescence, with an accordionlike dorsal fin that nearly doubles its height when flexed. An entire rainbow flashes in its scales—silvery blues and pinks and greens that slip away as soon as you try to fix words to them.

The inland Inupiat, I have read, once "cooked" grayling by slipping them in their boots as they walked; after an hour or two, the meat was considered ready to eat. We are not so hardy. Julia fillets the fish, and we fry it in powdered butter, salt, and pepper. At dinner, we pass around the piping white flesh: Soft and flaky, flavorful and rich, it is the best thing I have tasted in weeks.

* * *

June 24: This morning we inflate our rafts and begin downriver. Under placid skies, the water is a sheet of bending glass—sliding around beaches, crackling over gravel bars, pooling silent in deep spots. A northern harrier wheels overhead, and glossy swallows flit in and out of burrows along the banks. A furtive mammal—a martin or ermine—sneaks from hole to hole, hunting for eggs.

It seems improbable to find such an outpouring of life here at the so-called gateway to the New World, which has never offered inhabitants much more than a toehold on life. Sometime between fifteen thousand and thirty thousand years ago—before the last great ice sheets disappeared from North America, when the sea was hundreds of feet lower than today and the Bering Land Bridge connected Alaska to Siberia—the earliest Americans entered the continent, finding what the Canadian archaeologist Robert McGhee once called "the coldest, darkest and most barren regions ever inhabited by man." For at least ten thousand years, small bands of hunters roved these headwaters of the Noatak, but even as the climate softened and the ice retreated, life remained marginal: There was simply not enough solar energy.

"It is a matter of low productivity, of life stretched across vast spaces, but exquisitely adapted to survive," wrote the conservationist John Kauffmann, who explored these mountains in the 1970s before campaigning to create the Gates of the Arctic National Park. The twentieth-century Nunamiut hunter Simon Paneak put things more succinctly: "It is hungry country."

The key to survival here has always been to keep moving. It takes something like one hundred square miles to sustain a single grizzly here, and more for a pack of wolves. The caribou travel upward of two thousand miles in a year, both to find enough lichen and herbs to nourish themselves—these tundra plants, after all, grow less than an inch per decade—and to escape the wolves and life-sucking mosquitos that follow. Humans once followed, too. In the millennia before the introduction of firearms and motorized travel allowed them to settle year-round in Anuktuvuk Pass (some one hundred miles northeast of here), the Nunamiut, whose name means "caribou people," were considered perhaps the best hikers in all of Alaska, often traveling up to thirty miles in a day. The Nunamiut ate every part of the caribou, often raw: intestines and brains, nerves and tendons, the stomach and its contents, even newly grown horns. The fat white grubs that lodged themselves beneath the caribou's skin were considered a great delicacy.

Even so, this country is so lean that in the centuries before Western contact, all of Northwest Alaska—a region larger than Maine—held only about 5,000 people, and this number fluctuated with a volatility unknown to tropical and temperate regions. At the end of the nineteenth century, when the caribou herd plummeted—as it has always done from time to time—the human population fell to roughly one thousand. Some Nunamiuts fled northeast to hunt whales on the coast, where Ernest Leffingwell met them; others remained in these mountains, boiling bones and eating caribou skins, growing steadily weaker until many starved to death. I shudder at the thought of those cold, dark winters: Surrounded by the dead, the survivors could not even spare caribou fat for candles to light their windblown tents.

* * *

For two nights, Russell and Maddy and Julia and I camp at the mouth of a southerly tributary called the Kugrak, Iñupiaq for "old river," while Roman and Forrest search for trees in its headwaters. Two other small parties pass. (Upon hearing of my Achilles problem, one woman volunteers that she is a professional acupuncturist, and—would we believe it—has brought needles. Next thing I know, I am lying on a gravel bar, letting her perforate my leg.) Drawn by the scenery and fishing, scores of visitors float the Noatak every summer, but a few miles away—in the mountains where Roman and Forrest have gone, and which we will all enter in two days' time—lie valleys that may not have been traveled this century.

When Roman and Forrest return, we continue downriver through darkening skies. Small thunderheads shuttle, blurring distant peaks, projecting rainbows into holes of sun. By afternoon the horizon is a sheet of gray. In the flat light, the lush greens of the mountainsides turn neon; the yellow potentilla flowers and the fuchsia fireweed stalks pop wildly. The clear water seems lit up from within.

"It feels like we're in the American West before the droughts," calls Roman.

"Aye, Meriwether," replies Forrest.

The first veil of rain arrives, pelting our flotilla with bullet-sized drops. When it clears, we see brown gashes in the mountains—

permafrost slumps. From one of them comes gushing a black creek. It roars into the Noatak: pumping the current fast, billowing up arabesques of sparkling sediment beneath our boats, tearing at the fast-eroding banks, from which hang mats of disintegrating tundra. A willow shrub teeters over the boiling river, its roots exposed and fluttering.

* * *

June 28: My rest is over. Today the hiking begins, south into the Schwatka Mountains. The known tree line is on the south side of the Schwatkas, and Roman's new forest is still some fifty miles to the east, but six days ago, on the hike over from the Arrigetch, Julia spotted a seedling. Two days ago, Roman and Forrest found several more on their survey up the Kugrak. Perhaps the advancing front is wider than even Roman suspected. We will see.

The valley we climb is lidded by heavy clouds, which spit cold rain as we cross and recross a fast, clear creek. I focus on putting one foot in front of the other. My Achilles has not hurt in several days, but I have not really tested it.

This valley is called the Ipnelivik, Iñupiaq for "sheep raising place"—likely a reference to the steep, cliffy terrain. Aside from the Arrigetch, the Schwatkas comprise the highest and most rugged rock in the western Brooks Range. For millennia they have presented an impermeable barrier to the thick spruce forests that cover the south side of the range. This tree line is not just an ecological boundary, but a cultural one: It roughly divides the polar Iñupiaq (formerly categorized as "Eskimo") cultures from the woodland Athabaskan (formerly categorized as "Indian") ones, which extend in various forms down to Northern California. For the Koyukon Athabaskans, who reside just south of the Brooks Range, the white spruce is nearly as important as caribou are to the Nunamiut. "Almost every part has special uses and significance," wrote the anthropologist Richard Nelson in 1983, adding that the white spruce was—animals included—"probably the most important single species in their economy and lifeway."

While the traditional Nunamiut pitched caribou skin tents over willow-stick frames, the Koyukon built log cabins. While the Nunamiut almost exclusively walked, the Koyukon built a variety of wooden watercraft. They gathered needles to boil for medicines, roots from which to carve bowls and lash baskets, and pitch to disinfect cuts and caulk boat seams. The broad, dark canopy of the spruce sheltered winter travelers from wind, and also radiated solar energy downward, providing relatively warm camping. And of course, the wood provided an abundant source of fuel through the winter, a luxury not afforded the Nunamiut. In part for these reasons, the Koyukon believed that each of these trees possessed a powerful and benevolent spirit, which could be used to ward off dangerous ones. The white spruce, a Koyukon saying went, "will take care of you."

Nonetheless, the forest dwellers made summer voyages beyond the tree line to hunt caribou, just as the Inupiat traveled south of it for the same purpose, and much of our present navigational information we owe to both cultures. "They explored everything up here," Roman has told us. "The passes with Native Alaskan names are usually good, and the ones with English names usually aren't." The pass we are heading toward now is not named, but the creek on the other side is called Ulaneak, Iñupiaq for "wrong route" or "blind pass"; and though the footing is not bad here, already my Achilles is biting through the eight hundred milligrams of ibuprofen I took this morning. After a few hours, Forrest turns to me with a troubled look. "Your gimping has gotten noticeably worse," he says.

A few miles later, we make camp in the lee of a rocky knob that rises out of the valley, the only wind protection for miles. There is no floor on our tent, and the ground is wet; water pools in from the edges.

After dinner, while Julia is sticking my leg with her hiking pole à la David Cooper, she suddenly squeals; clenching my teeth in pain, I wonder if part of my leg has come off in her hands. But looking over my shoulder, I instead see a tiny puff of feathers: A merganser chick, perhaps two weeks out of the egg, has wandered out of the rain into the center of our tent. Noticing us turn, it emits a barely audible *cheep*, then pops up on toothpick legs and scampers out of the tent.

* * *

By morning the weather, nasty yesterday, has turned positively malevolent. The wind whines, lashing ropes of rain against our tents. "If it's like this here," says Forrest, coming over to our tent with Roman, "it's howling at the pass." Roman decides, for the first time in my travels with him, to stay put until conditions improve.

All day the sky unloads. For water, we simply leave our pots under the drip line of the tent; after twenty or so minutes they are full. We are camped in the overflow channel of a small creek, and the water rises so quickly that Roman and Forrest roll rocks to build a dam. All night, boulders rumble like thunder in the creek, waking us from time to time.

* * *

Early next morning the sky is still snotty and the snowline has descended low in the mountains, but the wind and rain have eased. We break camp and continue up the narrowing valley, under towering, sedimented scarps that rise into mist. At a small coin of a lake, framed by towering snow ridges, Roman suddenly stops, possessed: At his feet, protruding from a mat of grass and Eskimo potato, is a six-inch spruce seedling.

"This is the third pass we've come over along the Noatak, and all three of them have had spruce on the Noatak side!" he exclaims. "And they're all young and healthy. Just look at it!"

NORTH TO THE FUTURE

Indeed, it is greener than green. The distances between bud scars—the bulging knots on branches that separate each year's growth—are several inches long, more than impressive for an Arctic seedling. The question is, how did it get here?

We are still miles north of the previously documented tree line, and the pass in between is no place for plants, as the wind numbing our faces attests. But this south wind, Roman thinks, may hold the answer.

These mountains have no doubt always funneled strong winds, but winds from the north have historically prevailed in winter, when cold, dense air drains south through the passes. According to Native accounts gathered by the twentieth-century anthropologist Ernest Burch, these north winds (blowing north to south) sometimes blew so cold and hard in winter that they froze caribou to death. After storms, Natives sometimes hiked up into Howard Pass, forty miles north of here, to gather the carcasses—and it was this wind feature, rather than any river or mountain, that constituted the border between the Nunamiuts and another Iñupiaq nation called the Nuataagmiut. Sixty miles west of here, in the Hunt River valley, the same north winds were known to drive stones with such penetrating force that they literally mowed down trees.

These wind patterns must have changed in recent decades, or we would not find young trees blown *north* over the Schwatkas. The very limited wind data from the region bears this out, as does satellite imagery, which shows south-to-north snowdrifts. The reason for this is almost certainly the same reason the Arrigetch has seen squirrel-killing, bird-fostering snow years: the retreat of year-round sea ice from the Chukchi Sea. In October and November, when the temperature has dropped enough for precipitation to fall as snow but not enough to freeze over the ocean, North Pacific storms now send wet, windy air parcels into the western Brooks Range. Snow, reasons Roman, smooths over the uneven ground, and wind compresses the snow to a crust, paving the way for seeds and sending them skidding along for miles, up and over the passes.

This forest's migration, then, is not just a matter of warmth or water or nutrients, as we initially suspected: It depends also upon the dispersal of seeds. "In this great chain of causes and effects," wrote Humboldt, "no single fact can be considered in isolation."

* * *

The pass, when we reach it, is a wind-blasted gorge devoid of all but lichen. We descend through a surreal moonscape of truck-sized boulders, fallen from the mountains overhead. I lag behind the others, leaning heavily on a hiking pole Julia lent me, and by the time we reach the first spruce trees on the south side of the pass, my forehead is sweating, even as I shiver from the wet cold. Roman waits up to check on me, and in his face is the softest look I have seen from him on the trail.

The last few miles are interminable. Surrounded by steep, alder-hashed walls, there is no place to camp, and nowhere to walk but *in* Ulaneak Creek—a swollen torrent that sweeps Julia's pole out of my numb hand. Finally we come upon a raised bench large enough for three tents, but we've been out in cold rain for ten hours and are all fast approaching hypothermic. What we need, more even than rest, is fire.

Even in the heaviest rains, old spruce trees harbor dry boughs near the base. Using our unresponsive fingers like vises, we scrape together enough of these for Roman to coax a flame, and then drag waterlogged branches over to dry. The blaze grows so high that it evaporates the rain above us, radiating a hissing bubble of warmth. Around it we stand dumb, watching steam rise off our shivering bodies.

I want to collapse right here, but as my mind slowly wakes from stupor, I realize the meaning of Roman's look earlier: He thinks that even after flying me to Coldfoot to rest, I will have to fly out again with Forrest in a few days. That my journey is coming to an end. The thought is more piercing than the exhaustion or the pain, so I fill my pot with water and put it next to the fire.

Once it boils, I fill my water bottle and use the hot surface to massage my calf. Julia pitches the tent, and then sticks my tendon. After, I stretch, ice in the river, swallow four more tabs of ibuprofen, and finally—not bothering to change out of my grimy hiking clothes—drop to my sleeping pad. Within seconds I drift into a wooden sleep.

* * *

Over the final two days and thirty miles to the cache I retreat to a dull point somewhere deep in the brainstem. Each morning and night, I repeat my therapeutic ablutions with Julia's help. Each day I take two thousand milligrams of ibuprofen—"Vitamin I," Roman calls it—and still the pain is bad enough that I don't notice the blister forming on my heel until I remove my shoe at day's end and find it weeping. But the others are suffering now, too. The footing is hellish, and winds through dense groves of alder that dump leaf-fulls of water on us; Roman aptly compares it to an "endless car wash." In between showers, the bugs swarm like zombies, desperate for a blood meal before the rain returns.

"I've gotten better at just swallowing mosquitos," says Forrest one afternoon. "They never show this part in the movie."

The first mosquitos to appear, in early June, were of the *Culiseta* genus—large, clumsy pests that fall apart on contact. These are *Aedes*, a smaller and quicker genus that can survive a swat or two, somehow popping back out to continue their infernal hunt. Not even the Natives are inured to this pest. According to Koyukon cosmology, Raven—creator of the universe—introduced mosquitos to flush his six wives who had run off into the woods. The plan worked; no sooner did the bugs descend than the women came rushing back to him.

Eventually we are all haggard enough, and the boiling Ulaneak level enough, that we opt to inflate several boats for raft-packing. Roman suspects that we are the first to raft this river in any craft; "It's too splashy," he reasons. We manage to stay upright through the rapids, and on the afternoon of July 2 reach the Ambler River. Then we begin the final seven-mile hike upriver to our barrels—a silent, rain-numbed procession.

Sometime after solar midnight, we find the cache on a gravel island and make camp. No sooner have we finished dinner than the sky darkens to true midnight. Moments later a gust of wind blows out the front flap of our tent, which starts luffing like a loose sail. Rain falls like artillery fire, and—finding no place to drain on the disappearing island—filters in and around our sleeping quilts.

Yelling to hear one another, taking turns holding down the tent, Julia and I throw gear into our packs, pick up our tent like a beach umbrella, and begin looking for a path to solid ground. Roman and Forrest are scrambling to gather gear from beneath their flattened tent, which apparently blew down in the wind. Maddy and Russell are already packed and probing the river.

Soon they find a channel shallow enough to ford. We cross to the far bank, and then huddle under a large spruce tree until the squall dies. As Julia and I pitch our tent, Roman approaches.

"You look like a new man," he says. I don't, but my Achilles has not gotten *worse* in the last two days, and the others are trashed now, too.

"Thanks to the Advil," I reply.

"Whatever it is, it's working."

Twenty minutes later, as I write these notes—at 2:50 a.m. on July 3, my twenty-fourth birthday—I'm glowing inside: My journey is not over yet.

4

"Cut stuff," says Roman the next morning. "Go as light this next leg as you possibly can."

To reach the heart of the new forest, we will need to carry ten days' worth of food over a pass that Roman expects to be no less trying than the last. What's more, Forrest flies out the day after tomorrow, and we all will have to take on more group gear.

The trouble is, I don't see much to cut. My base weight—which includes a stove, tent, sleeping quilt, sleeping pad, a pair of hiking clothes, a pair of sleep clothes, fleece, down jacket, rain gear, science equipment, rafting equipment, Leatherman, blister pads, toothbrush and paste, notebooks, a Kindle, and a few other personal effects—is less than twenty-five pounds. Ten days of food and fuel is another twenty-three. Just south of fifty pounds for a ten-day, multisport Arctic science expedition strikes me as spartan, but I can't say this to a man who has traveled hundreds of miles with little more than a lighter and the shirt on his back.

So I begin with the easy stuff, cutting my water treatment drops—which I have not yet needed—and my point-and-shoot camera and batteries. Then I move to the food pile, trimming it by roughly a third. Finally I pick up my mini Bluetooth keyboard, which weighs less than a pound and allows me to quickly transcribe handwritten notes into my phone each evening. It's worth at least five pounds of chocolate to me, but Roman has been eyeing this piece since I arrived in Alaska, and is watching now with no uncertain feeling. "Come on!" he finally cajoles, and I release the keyboard, regretting it instantly.

* * *

The next day we conduct fieldwork near the barrels, and the day after, we hear the faint whine of an engine. Soon a small plane appears, banking into the Ambler valley and landing on the river island. But when we wade out to send Forrest off, we instead find four jean-clad software engineers from Silicon Valley. In Alaska for Fourth of July weekend, they explain, they decided to catch a "flightseeing" tour up to the Arctic. One is employed at Google, another at Palantir, and a third says he works for "The Facebook company," as though we might have missed that bit of world history, like some filthy horde of latter-day Rip Van Winkles. We, in turn, ask them to look up the weather report, as though a cloud of Wi-Fi might cling to their spotless puffies. After about half an hour on the ground, they pack up to fly back to Fairbanks, but before going they ask to take a picture with us. It occurs to me only now just how putrid we must smell.

After, as we sit around a fire, talk turns to the so-called Ambler Access Project, which may soon bring more than flight seers and scientists. A year ago, the Trump administration permitted a 211-mile mining road that would thread the southern edge of the Brooks Range, from the Dalton Highway to the mouth of this valley. Upon entering office, President Biden promptly ordered a more thorough environmental review, beginning a political and legal struggle that continues today. Proponents argue that mines in the region will bolster Alaska's faltering economy while yielding copper and zinc, minerals required for a clean energy revolution. Critics argue that the road—which will cost something like $1 billion and require building an estimated forty-eight bridges and nearly three thousand culverts—is an ecological nightmare, and that more substantial mineral reserves can be found outside of America's greatest remaining wilderness.

Very mention of the road sets Roman fuming. In a legal declaration this spring, he argued that the Brooks Range could "absorb the climate damage we humans are inflicting," but only if we give it a chance:

> Forty years ago the Alaska Pipeline was completed, cutting the Brooks Range wilderness in half. This Ambler Road will take it off at the knees. The oil development in the NPRA and the Arctic Refuge will lop off the head... someone must speak for the land and its creatures. I'm offering up something selfish but non-consumptive. Something that's more important than hope. A future for a place that's intact now. To mar it for a decade's worth of dollars is a sin in my religion.

* * *

A few hours later, a second plane arrives. We hug Forrest goodbye as two other men—more guests of Roman's—pile out. The first is a sturdy, straw-haired scientist named Logan Berner, who has led advances in Arctic greening research—cobbling together the clearest satellite imagery available and using advanced computational methods to estimate trends across the Arctic. Close behind is his colleague Patrick "Pat"

Burns, a lean, redheaded man in his early thirties. Their lab at Northern Arizona University is funded by a NASA grant to link remote sensing studies with field science to improve models of Arctic climate change, and they've found an ideal collaborator in Roman.

Both have spent some time on the ground, too. Logan grew up in Southeast Alaska and came to remote sensing from fieldwork, while Pat rafted the entire Grand Canyon last November. But their packs are large, and Roman wastes little time in asking—politely—to gut them.

"We share everything," he explains. "There are five of us, and we have one mini knife. We have one nail clipper."

"One piece of toilet paper?" jokes Logan.

"They don't have *any*," replies Roman, gesturing to the rest of us. "I've got them all trained up, so ask them questions."

We help them trim their gear, and then all pile into Roman's tent for a dinner of moose chili that Roman's wife, Peggy, made for us, along with olive ciabatta and baklava from a bakery in Flagstaff. For dessert, we eat Peggy's brownies, which cooled en route to the Anchorage airport this morning. As our stomachs distend, Roman describes our findings from the past month: the arboreal colonization of the Arrigetch, the young trees spilling over passes in the upper Noatak, the seedlings scaling the walls of this valley.

"It's almost hard to believe," says Roman. "You don't *want* to believe it. But then you look around, and there are seedlings everywhere."

* * *

Next day, July 6, the seven of us continue north along the Ambler, turning west up a tributary that is not named on any of our maps. The river climbs above the forest into arid mountains laced with seams of iron. Springs seep out of the barren slopes, and around these grow lush fans of moss and pipe-cleaner spruce, like desert oases. Down below, the riverbed is coated in white-yellow sediment, perhaps calcium carbonate leached from limestone. The water—a color I can only describe as "Windex blue"—is disconcertingly beautiful.

The valley constricts to a gorge, forcing us down along the stone riverbed. Russell—the largest, strongest member of our team—is carrying the heaviest pack, and a nubbin of rock snaps under his weight; he comes crashing down on one knee. Blood seeps through his pants, and the pain makes him nauseous. Concerned that he may have broken something, we make camp on a flat bench across the river.

* * *

July 7: By morning, Russell concludes that his knee injury is manageable. We collect samples of several trees and shrubs, and then climb west, up to the last, unnamed pass before the Noatak River Basin. Wind whistles, rain falls. I am pixel walking, and there is nothing to record but rock and lichen—a good thing, since my fingers are soon numb from the wet cold. For a time it appears that the scree slopes we follow may tail off into the roiling gorge below, forcing us to turn back, but a few thread-like trails tell of caribou that have passed through here, and Roman trusts them almost as much as he trusts the Natives.

His trust is well placed. After several hours, the terrain levels out into a broad, windswept corridor shrouded in clouds. Sparse mats of moss and dryas cling to life among the rock. Several hundred yards beyond the pass, a single line gives verticality to the waste: the bare skeleton of a spruce. *What forces gave you life here?* I wonder. *What took that life away?*

Just below the pass, along the first trickles of the Imelyak (Iñupiaq for "little water"), which flows northwest to the Noatak, grow several healthy seedlings. Beyond them, we come to a verdant specimen more than six feet tall.

"Look how fast it grew," says Roman. He grabs the bright green tip of a branch, which extends several inches beyond the last bud scar. "One year of growth right here. Isn't that crazy?"

We did not intend to core more trees today, but Roman wants to prove his point, so I dig through my pack for the appropriate tool, hands clumsy with cold. Logan blows on his own for warmth, then takes the T-shaped instrument and begins screwing the long end into the tree's base.

"I wouldn't be surprised if it's only thirty years old," says Roman. It seems an absurdly low guess; even among established forests, it is common to find thirty-year-old trees that are scarcely a foot tall.

But Roman's guess is high.

"Whoa," says Logan, pulling out the core and counting its rings. "These are huge." He looks up. "I only count twenty."

We all peer over Logan's shoulder.

It is strange to look at the rings of a tree and see the years of your life collapsed to a bit of wood. It is stranger still in the Arctic, where the years are reduced to slivers, scarcely discernible to the naked eye. But these rings are wide. I count back—to junior year of high school, when I first read Thoreau. To seventh grade, when my friends had growth spurts and left me in the dust. To second grade, when my family moved to Boston for a year, and the Red Sox broke the Curse. At pre-kindergarten I hit the pith.

"Twenty," confirms Roman, stepping back. "Twenty…"

"So it's probably my age," I say, adding four to account for the years before the seedling reached the height of Logan's core.

"And it's taller," Roman replies, almost to himself. "A tree in the Arctic is growing faster than a human."

"Phenomenal," says Logan.

Soaked and shivering, we continue to descend, but keep finding bigger and older trees—a sign that this population did not travel over the pass we just hiked, but came from somewhere else. After a few more miles we make camp beside the Imelyak, and then all seven of us pile into Roman's floorless tent, curling around wet shrubs and each other's limbs. Faces thawing over cocoa and tea, we debrief.

"I think the trees blew up another river," says Roman. "And when those first trees got old enough to reproduce, updrafts sent their seeds up here. That's how you get Ben's twin." He grins at me. "But they didn't start here. I'm pretty sure we're going to find bigger trees farther west."

"So," begins Logan. We are so crammed that I can feel his voice vibrating through my leg. "Now it's time for the Maury Povich question: 'Who is the baby's daddy?'"

* * *

July 8: Cold rains continue as we cut west across the rumpled foothills of the Schwatka Mountains, so green they are almost yellow. In the flat light, fields of scouring rushes radiate freshness, their jointed design unaltered for half a billion years. The swales are flooded with the sweet smell of rainfall and thawing soil, and in one such glade the water flows along with our feet, growing swifter and louder until we come to the base of a knoll and find a gasping hole in the earth. Into its black maw the water rushes and sucks, echoing back up from somewhere deeper than memory. We give this shortcut to the underworld a wide berth.

The trees grow larger and more numerous as we continue west toward the lower passes through the Schwatkas. We stop periodically to record their dimensions, measure recent growth, and take cores. In a stable climate, a tree's outermost rings are thinnest, because new growth is distributed over a greater area. Not so here. If anything, the most recent rings are wider than the rest—so wide that they are almost translucent, and look less like wood than gel. It's a clear sign of rapidly accelerating growth.

So *why?* The wind may explain how seeds got here, but not how they germinated. Nor why they're growing as if they lived hundreds of miles south of the tree line, while forests in the eastern Brooks Range remain stagnant. As we walk, Roman and Logan theorize.

"Here in the West," says Roman, "the Chukchi Sea is now sending a bunch of snow in October. That blankets the land, so it's insulated before January's super cold comes. And then the microbes in the soil, they actually metabolize all winter long if they're warm enough. They can be breaking down organic matter into nutrients and fixing nitrogen."

"The snows would also protect seedlings from wind and freeze-thaw events," adds Logan.

"Yup," says Roman. "And then in spring, the seeds have water before the sun comes out and cooks 'em."

He pauses for a moment, then chuckles. "Back in 2019, I thought it was all kind of simple. But it's actually pretty complicated. But then again, it's pretty simple: A place that's warm, wet, and nutrient-rich is good for trees. Imagine that?"

* * *

In the afternoon of July 9 we come through foothills into a broad, lush valley named Amakomanak. Across the river, several miles away, green spires rise on the slopes—and judging by their position and size, they may well be the original colonists of the Noatak River Basin.

It strikes me as almost unbelievable that no one has seen these trees before. True, the Brooks Range is vast; in 1958, Clarence Rhode, the Alaska director of the US Fish and Wildlife Service, disappeared while flying two others in the eastern Brooks Range, and the largest air and ground search in Alaskan history to that point failed to find them. (It was not until 1979 that two hikers came across their wreckage in a mountain gully.) But archaeologists have been through this valley at least twice in the last thirty years to document arrowheads, carved roughly 7,500 years ago by peoples who preceded the Inupiat, and several other parties have been through here recently, too. Since first finding spruce in the area two summers ago, Roman has contacted everyone he can think of, and all told him the same thing: *No trees. I would have noticed.* After all, how does one overlook a full-grown tree towering over the Arctic tundra?

But it is easy to overlook the things that do not speak or flash, even where they define the boundaries of life. "It's like when you're walking through a field of flowers," says Roman, "and somebody goes, 'Wow, look at this yellow flower!' And you reply, 'Oh, that's beautiful!' And then you realize you've been walking through it for hours. You just never noticed."

As we begin across the valley, the sun emerges for the first time in nearly a week; its warmth is heavenly. We make camp on a gravel wash and lay everything that is wet—which is everything we have—out to dry. Pat produces a deck of waterlogged cards, and Roman dries them in his cook pot, tossing them over the stove like stir-fry. Then we play Up and Down the River and laugh as we have not laughed in days.

There is a giddy energy in camp tonight: We are closing in on the heart of the new forest.

*   *   *

Next morning we cross the river and slosh toward the trees. Mosquitos swarm out of the hummocks; Julia slaps fourteen on Logan's head. Reaching the forested slopes, we begin roving in groups, documenting spruce one at a time. Julia and I can hear Roman frothing from across the slope: "Look at this little shorty with all the cones! Look at that guy."

Toward noon, several of us converge on a particularly tall tree—the tallest, it seems, in the valley. Pulling a rangefinder from his pack, Roman backs up to measure its height. Not since the night he read his book aloud to us have I seen him so agitated.

"God, this one is a giant!" he says. "It's really loaded with cones. And it's so tall I can barely see 'em!"

The base of the tree is obscured by shrubs, so Pat measures it by hand, placing his pack where his reading stops. Roman gets quiet and measures from the top of the pack to the tip of the tree.

Finally he looks up from the rangefinder.

"8.90 meters to the bag," he says, voice trembling. "Pat?"

"1.37 meters to the base."

Roman is a math professor, and one of the quickest minds I have encountered. But now he pulls out his phone to check his arithmetic. "So that's...

"That's...

"That's...10.27 meters. God. That's giant. God, I'm kind of overwhelmed right now."

He takes a moment to collect himself, then turns to us. "I've been from one end of the Brooks Range to the other, and I haven't seen anything this dramatic."

* * *

The next day, July 11, we hike west to a clear river flowing out of the Schwatkas called the Cutler, where Roman first discovered trees on satellite imagery in 2019. The day after we follow the Cutler toward our next food cache, some twenty miles northwest as the raven flies. Gradually the foothills of the Schwatkas recede into the broad basin of the

NORTH TO THE FUTURE

Noatak, and still we continue to find spruce—ten, twelve, fifteen miles beyond the known tree line. It is a sight to watch Roman's mixture of glee and dread as we pick among the trees. He's like a storm chaser, watching the world's first category 6 hurricane whirl toward shore.

For my own part, the implications feel too large, too uncertain, to wrap my head around. The trees, after all, are still; the tundra is silent; and hard as I look, I find in them no indication of how to feel. Mostly I am uncomfortable. The skimpiness of my two-thirds rations has compounded into a constant, gnawing ache, and I am hungry all the time.

And yet those implications, when I can muster the energy to consider them, are staggering. Now that trees have come over the Schwatkas, they've cleared the last great topographic barrier to the Arctic Ocean, more than two hundred miles north. Spruce forests reproduce irregularly, in bursts called masts that typically happen once every two to twelve years, but both of the last two summers were mast years in the western Brooks Range—yet another indicator that these forests may be on the verge of a population boom. Already, down among the grasses and shrubs, invisible except on hands and knees, a new wave of seedlings has taken root. There is no telling where seeds have blown, or where they will go next.

What new forests like this one mean for Earth's climate remains wildly unclear. Most evidence suggests that the net result will be a strong warming effect, caused primarily by increased sunlight absorption in spring and fall, when dark spruce canopies protrude above snow. But just how much sunlight will new forest canopies absorb across different seasons, and how will this interact with changing snow patterns? How large will these trees, and their counterparts farther south, grow, and how much carbon dioxide will they turn into wood and carbon-rich soil? What effect will they have on the permafrost, which has apparently already begun to thaw and stores many times more carbon than these trees ever could?[1]

"It's still really an open question," said Colgate professor Mike Loranty, one of the world's leading experts on the interactions between boreal forests and climate, when we spoke a couple months before the trip. "This warming, it's catalyzed a whole host of ecosystem changes that are hard to tease apart, and that make it hard to determine causality. A lot of what we're seeing right now are transient effects, because the system hasn't reached a new equilibrium. It's just really unclear how it's all going to net out."

Another way to approach the question is to consider what Earth's climate was like the last time trees grew this far north. Scientists have found spruce fossils along the Noatak River, and Inupiat have found sizable logs protruding from the permafrosted banks of the Colville River, on the North Slope. All such fossils scientists have managed to recover are "carbon dead," which means they date back to at least the last interglacial period more than a hundred thousand years ago. Back then, global climate was several degrees warmer, and the Chukchi Sea was likely ice-free year-round. Sea level was twenty to thirty feet higher than today.

"We can't rule out the possibility that there were trees here since then," says Roman as we walk. "But if that were the case, I would expect to find a population somewhere that's really old. Or you'd find some stumps, since things are so slow to decompose up here. And I haven't been everywhere, but I've been looking, and I haven't seen any.

---

1 This last question remains particularly fraught. In summer, trees cool the ground by shading it, an effect that has been proven through tree-removal experiments. But in winter they trap snow, forming drifts that insulate the soil from the deepest cold. On the flip side, a few recent studies, conducted on shrubs, suggest that trunks can serve as conductors, pumping heat *out* of the ground in winter. The relative weights of these effects are still murky.

NORTH TO THE FUTURE

Of course, you can't say that in a paper—'I've looked around a lot.' That doesn't really count as scientific evidence. But that's what I think."

"What we know for sure is that trees don't show up here withoughh—" Logan coughs on a mosquito, spits, and continues. "—without a reason. It's a bellwether. Change is afoot."

* * *

July 13: A high-pressure system has settled over the Noatak River Basin, and the air grows heavy and hot; the sky is an open dome of exposure. We pass a dozing grizzly who does not even stand at the sight of us. I wouldn't either, if I were glued inside a shag carpet. In the balmy stillness, bugs reign across the tundra, swarming in such numbers that they shimmer the horizon, their whines coalescing into a deep, engine-like hum. I slap nineteen mosquitos on Logan's head, and later four horseflies on my arm.

At least our tents have bug nets. The Natives had only "smudges"— bits of old wood or birch fungus that they burned continuously, to limited effect. How they slept at all in midsummer, I cannot imagine. The bugs are so bad some evenings that we do not leave the tents even to relieve ourselves. "We dug a sewage system in there last night," says Russell this morning, as we slog through tussocks.

"The trees are cool," adds Maddy. "All of the rest of this I'd kind of like to forget."

But the trees continue. On the afternoon of July 14 we reach our rendezvous point with the pilot, a gravel island in the Cutler River, and happen to find several seedlings right here. Already, we've determined that this is the fastest-moving tree line known anywhere on Earth. But just how far north does it go?

5

July 15: Today, Logan and Pat fly out, and Roman will go with them; he has family matters to attend to in Anchorage. Waiting for the plane, he briefs us on the coming leg.

"It's a treasure hunt," he says. "We wanna know how far the trees go. I don't think you'll find any up on the Noatak, but look anyway."

Roman's former student Allen Dahl, who injured his own Achilles on last year's expedition, will arrive on today's plane, and the five of us will raft the Cutler fifteen miles north to the Noatak, and from there another two hundred miles west down the Noatak to our next resupply. The plane will also bring thirteen days' worth of food, which should give us plenty of time to hunt for trees, pixel walk ten kilometers each day, and gather shrub samples.

"Russell's the boss," Roman adds. "No *Lord of the Flies* shit. We gotta get this done."

As he finishes addressing the last of the objectives he's scribbled down for us, a plane engine hums into earshot. "Good timing," he says, handing the list to Russell along with the satellite phone. "I'm taking my gun, so you guys only have one now. If a bear comes swimming after you when you split up, break your paddle in half and just fuckin' hit him on the nose as hard as you can."

The plane's landing triggers a flurry of action—unloading packrafts and paddles, grabbing food, swapping out new equipment for old. The barrels hold my only change of clothing for the trip, and after finding it I duck behind a shrub to wriggle out and in. Minutes later we reseal the barrels, hug Roman and Pat and Logan goodbye, and stand back as the engine judders to life. I'm still rifling through my things when the plane starts rumbling forward, and Roman throws a peace sign from the cockpit.

"Guess it's just us now," says Allen, as the plane drones out of earshot.

I barely noticed Allen in the commotion. Lean and graceful, with green eyes and long, dark hair, he grew up in rural Alaska, born to a white father and a Yup'ik mother. He is a quiet presence; I met him two years ago, over beers with Russell and several other young men, and heard him utter only a few words. It says something, though, that Roman paid Allen's way here, even though Allen tore his ACL in a backcountry skiing accident this spring and will be of limited help with fieldwork. "He thinks before he speaks," Russell told me. Julia has called him "observant."

I reintroduce myself, and the five of us all pile into Russell and Maddy's tent to gorge ourselves on the new rations and play rummy with the crusty cards Pat left us. It is afternoon, and we will not leave until morning. For the first time this summer, it seems that time, food, and weather are all on our side.

The winds have blown most of the bugs from this island, and after cards I walk up to its northern tip, where the river has ground rocks and gravel into fine white sand. I sit and try to clear my mind. Since entering the Schwatkas two weeks ago, I have used every spare moment to treat my Achilles, and now, as I unclench for the first time in many days, I feel swept along by a current I can't see or name. And soon the trapdoor opens—the knowledge that everything and everyone I love will one day slip away from me, into the onward rush of things.

Wind blows cold into the wetness on my cheeks, and I try to focus on this sensation. The feelings of heat and pressure and tingling in my body intensify as they grow clearer, until they are everywhere. After a while I am the crisp on my sunburnt nose; the tickle of stray mosquitos on my skin; the weariness in my muscles. Gradually the senses open further—to the soughing of the willow leaves and the blood pumping through my body. The branches of light flashing up from the river. The sound of the river is deafening. The wind blows through.

* * *

Morning comes bright and clear, with warm gusts from the south. We break camp and set off onto the river in three red rafts: Allen takes one, Russell and Maddy share another, and Julia and I are in the third. The Cutler runs wide and swift now and slingshots noiselessly around bends, magnifying the smooth stones below. Each time a rapids begins to draw on our raft, my chest swells and I cannot help but grin. This buttery glide is what flying must feel like.

Grayling swirl in clear pools, and around most bends appear broods of mergansers and loons and greater white-fronted geese, the last of whose faces look dipped in milk. Julia has been teaching me to identify them. The birds here are not accustomed to visitors, and when they see us coming, the mothers herd their chicks not away from the river, but rather down its bank, squawking and ducking their heads in growing panic. We can do nothing to dispel the illusion that we are pursuing.

NORTH TO THE FUTURE

Every few miles we pull over to search the banks for spruce. We all have a pretty good sense now for where we might find them; "this feels sprucey," one of us will say, noticing a raised floodplain, or an apron between two habitats, or some other cluster of associations. But we find fewer as we continue north, and by the time we reach the broad churn of the Noatak River, we have seen none for several miles.

Ferrying across to the north bank, we haul out our boats and search again. It seems a good spot to look; the soil is thick and dry enough for moss to grow, and we have learned from Logan that a fire swept through this headland not long ago. Tundra fires are growing more common, and they clear the ground and prime the soil, making new habitat for shrubs and trees. But we are more than twenty miles beyond the tree line now, and finding spruce anywhere seems unlikely.

"Maybe it's too windy here," muses Russell after an hour or so. "We could look forever and find nothing."

But we know that the farther we go down the Noatak, the less likely we are to happen upon spruce. From here the river bends northwest, deeper into the tundra and farther from seed-bearing parents on the upper Cutler and Amakomanak Rivers. Russell decides we should camp here. In the morning, Julia and I will continue the search on foot, while the others float downriver.

Carrying our rafts out of the wind, we build a fire and then jump off a bluff into a clear pool. The water is bracingly cold, and we sputter and scrub for a few minutes before shivering out to warm by the fire, cook dinner, and crawl into our sleeping bags. The south winds continue to blow.

*　*　*

In the morning dark clouds sweep the sky; a rainbow bridges the river. After boiling breakfast in our tent, Julia and I hand over our things—all but a packraft, paddles, a tent, a stove, a bit of food, sleeping pads, and a single sleeping bag—to the others, who are stuffing equipment into their rafts. Russell hands me the gun. "Put on your spruce goggles," he says, "but no tunnel vision."

Up on the bluffs above the river, the wind gusts, bowing the tundra in shimmering waves. Julia pixel walks, and I pick our path through the muck-welled tussocks, letting my mind settle on the angle of wind on my neck, and the slosh of our wet feet, and the cool metal of the gun in my hand. From time to time I look up to take my bearings and find the whole land rippling for hundreds of square miles, as far as the clouds permit vision.

Behind us, a cloud of dust rises from the Cutler like a desert highway. Funneled by the riverbed, the wind seems to be sweeping fine sand and glacial till into the sky, and probably seeds, too. We see no sign of trees, but the dust gives Julia an idea. "Maybe it's just too windy here," she says, echoing Russell's doubts from yesterday. "Maybe we need to find someplace more protected."

After several miles, we come to a curve in the headland, where an old bend of the Noatak has carved a thousand-acre floodplain. The plain is shielded from the south wind; only a slight breeze swirls in from the east.

"Like this," says Julia. "It's sheltered, and there's a bit of a wind eddy, where a seed might settle."

We descend the crumbling bluff to the plain, where dryas and willows have begun to grow. And there it is: two feet tall and green as hell. Roughly twelve years old, and twenty-five miles north of the tree line.

We howl in glee.

*   *   *

In camp tonight, the five of us set up our two tents facing one another and break out a bottle of bourbon we smuggled into the barrels when

Roman looked away. Raising our cups, we toast to this little spruce, living miles beyond precedent. Thriving in this wild, changing land.

"Compared to the East," says Julia, "everything just seems so active here. So fast. Even in relation to our own lifetimes."

"Our parents were growing up, and having us," adds Russell. "We were all just living our lives. And meanwhile these trees have been out here, making history."

* * *

The Noatak takes us northwest, deeper into tundra. We continue to look for seedlings but find none. Each day, a pair of us—either Maddy and Russell or Julia and I—set out on foot to pixel walk for ten kilometers, after which we inflate a raft and float, meeting the other three at an agreed-upon bend in the river.

This schedule is undemanding, especially for those who don't walk, and during long hours on the river I find myself growing restless again, turning inward. Allen, meanwhile, seems entirely at ease. I still have little sense for him; he sleeps in Russell and Maddy's tent and does not speak much; and on the river, Julia and I occasionally lose sight of him and pull over to wait, only to find that he has stopped paddling and is staring up at the clouds or bluffs, a lock of hair in his mouth. Less urgency I have rarely seen in a person, and the quality grates me. *Has he forgotten that we are here?* I think. *Doesn't he see us paddling?*

I look to Julia for signs of irritation, but she seems comfortable sitting still, too. And the truth is there is no good reason not to. The current is fast, and we have plenty of time to get where we are going. On the northwestern horizon rise the first gentle undulations of the De Long Mountains, named after the American naval explorer George W. De Long, who called this "a glorious country to learn patience in."

The Noatak meanders here, in places taking ten circuitous miles to travel two. Its course is constrained by the snaking moraines of glaciers past, which during the last ice age poured down from the Schwatkas and the De Longs, damming the river and forming a great lake. This lake grew and receded with the glaciers, occasionally overflowing the basin and spilling north toward the Arctic Ocean. Fifteen thousand years ago, we would be under more than a hundred feet of water here; twenty thousand more, and we would be under another hundred feet of floating ice.

These riverbanks are also DNA vaults. Rising a hundred feet tall in places, they are imposing walls of scalloped ice chunks, old stones, and ancient sediments, which Julia says have a peculiar, musty smell. Sometimes banks like these—eroded ever deeper by the river—expose prehistoric creatures: camels and sloths, mammoths and mastodons, tigers and the giant short-faced bear, who dwarfed the modern grizzly. Exquisitely preserved by permafrost, these animals come, in one paleontologist's words, "as close to immortality as the earth can offer." Occasionally stomachs still contain last suppers. Teeth hold thirty-thousand-year-old morsels.

Finding such specimens, the traditional Inupiat assumed that they lived underground and died after mistakenly burrowing into the light of day. This may partially explain the fantastical monsters that populate the Nunamiut storytelling tradition—enormous man-eating mice, fire-breathing demons with ear-to-ear mouths, and flying dwarves who wore anoraks sewn from caribou ears. But these stories are not so much stranger than the one scientists have pieced together. Something like three-quarters of North America's megafauna died off when the last ice age ended about twelve thousand years ago, and the portion was likely greater in Alaska. The entire food chain was reshuffled beyond recognition.

One explanation for this Late Pleistocene extinction is that humans, who had recently crossed over from Asia via the Bering Land Bridge, killed off the New World creatures. But in Alaska, much evidence suggests that changing vegetation played a decisive role. When the glaciers and ice sheets suddenly receded, the ocean rose, flooding the Bering Land Bridge and bringing maritime weather. According to sediment records, precipitation multiplied within decades, ushering in first grasslands, and later tussock tundra. The grasslands would have attracted a booming population of plains animals, like bison and those that preyed upon them. But when the implacable tussocks took hold, only a few creatures—like the caribou, with its large, spade-like feet and hyper-efficient gait—managed to adapt.

I spend free hours on the river reading science, natural history, and anthropology papers on my Kindle, trying to imagine how this land might have looked thousands of years ago. Trying to shake my default assumption that the world has always been as it appears to me now.

* * *

July 21: Russell and Maddy part ways with us this morning, heading south on foot to search for spruce trees. They will travel light and fast, circling back to meet us fifty miles downriver in three days.

Allen moves into Julia's and my tent, and the three of us continue in rafts. This new arrangement is cramped but offers occasion to get to know Allen. He grew up in Bethel, a six-thousand-person village in southwestern Alaska where many still lead a partially subsistence lifestyle. As a child he attended the federally funded Yup'ik school, where his mother taught the native language. He speaks slowly and carefully, with a quiet intelligence that does not clamor for recognition. I feel I do not need to point things out to him, because I know he is already watching, listening.

The river takes us southwest now, toward the edge of the Noatak

Basin and into the purple foothills of the De Longs. In the afternoon we come to a grove of balsam poplars huddled around a bright creek that rushes in from the north. "Cottonwoods," as most people here refer to poplars, are of less scientific interest than spruce because they tend to grow only along waterways. But they are the first trees we have seen in days, and the sound of their leaves swishing in the breeze holds us for a while.

Near the cottonwoods, perched on a shelf above the river, is a small, boarded-up cabin, the first we have seen in the Noatak River Basin. No one lives this far upriver anymore; the cabin is likely a Native camp, used seasonally for hunting and fishing, grandfathered in before the Noatak National Preserve was established in 1978. But I have read that just across the river, there was once a settlement of four dwellings. These were the seasonal homes of the Nuataagmiut, an Iñupiaq nation of roughly five hundred who once populated much of the Noatak Basin west of the Nunamiut. And before them, other peoples no doubt dwelled along these riverbanks, too.

We are nearing the Bering Strait, and every Native people in the Americas, from the Aztecs of Mexico to the Yaghan of Tierra del Fuego, must first have passed through this region from Siberia—on foot, or in boats along the coast. The sea has covered the old coastal sites, and even the most recent inland dwellings have been obliterated by flooding or swept away by the shifting courses of rivers. All that remains in the mountains are stray hunting artifacts: a scatter of chert arrowheads, a few lumps of ancient charcoal, a long-cooled stone hearth.

The exception is at Onion Portage, a place just south of the Schwatkas where a large bend in the river reliably funnels animals, drawing hunters during fall and spring migrations. A sandy knoll above the site has steadily deposited layers of sediment over the artifacts, preserving them and allowing archaeologists to place them in time. From this site, we know that humans have inhabited Northwest Alaska for at least ten thousand years. But for every question the record answers, it opens several more. At least three distinct cultures have inhabited Onion Portage, and there is a conspicuous hole in the record from about 4000 to 2000 BCE. No one knows why the site was abandoned then.

One theory holds that the boreal forest—which, spreading northward after the last ice age, was finally reaching the southern flank of the Brooks Range around that time—disrupted caribou migration patterns so radically that humans perished or were forced to migrate themselves. The twentieth-century archaeologist Don Dumond suggested that the region's next inhabitants were from a more southerly culture, which he called the "people of the spreading forest." But the true reason could be entirely other. The Arctic ecosystem is inherently unstable; the food chain is probably still finding a new equilibrium since the post–ice age extinctions. Composed of few species relative to more temperate ecosystems, its populations fluctuate wildly from decade to decade, and century to century.

"The Koyukon people live in a land where change is the norm and where stability is almost unheard of," wrote the anthropologist Richard Nelson in 1983. "They have seen the animals vacillate between proliferation and scarcity, each in its turn, cycle after cycle. They have learned the variegated mosaic of the land, and they have seen it move, a kaleidoscope of time."

* * *

We camp beneath the cottonwoods, and in the morning continue downriver into rising mountains. At gentle rapids, we aim for the swelling tongues that suck smooth through the center. The gravel riverbed turns here to bedrock—black, with veins of quartz that reflect the sun. I watch miles of stone pass beneath our raft.

Life has gathered in this section of the river. Nesting peregrine falcons shriek and dive from the bluffs, protecting chicks tucked among the rocks. Pacific loons take off at the sight of us, their feet pattering along the water's surface as they muster the speed for flight. Below our boats, fat grayling swirl, and salmon have begun to run. A lean gray wolf trots along the riverbank.

In the afternoon I notice a large brown figure on a gravel bar, lying so still that I at first mistake it for a rock. Soon it resolves into an enormous bull caribou, with a great crown of velvet antlers. I expect him to run at the sight of us, but the current carries Julia and me within a dozen feet—close enough to hear his huffing and see the steam coming off his nose—before he rises unsteadily, unfolding a thin, knobby frame. There is blood on his hind quarter, and fear in his eyes. I think again of the wolf from earlier.

In the evening we make camp on a gravel island and wash ourselves in the river. Then we build a fire and watch the smoke rise, and the water pass, and the tundra bow in the wind. After several hours Allen breaks the silence.

"Does anyone know what time it is?" he asks.

"No," replies Julia.

"No," I say.

"Good," says Allen, and we return to our watching.

* * *

July 23: We meet Russell and Maddy this evening at the mouth of a creek called Aklummayuak, Iñupiaq for "bear slope." They have found nothing but tussocks and mosquitos and are exhausted. We spend the night here, finishing the bourbon and laughing deliriously, drunk on inside jokes told so many times, and in so many different permutations, that they become funny again.

NORTH TO THE FUTURE

In the morning we continue downriver to the end of the Noatak River Basin, into funneling mountains. The river swings beneath cliffs of swirling rock; falcons wheel overhead. Soon we enter the so-called Grand Canyon of the Noatak, which in winter is said to be the coldest place in all of Northwest Alaska. The frigid air that pools in the Noatak River Basin spills through the canyon, adding windchill to temperatures that are already as low as anywhere else in the state. In the days before snowmachines, Natives fed their sled dogs particularly well here, because otherwise they were known to freeze to death, even while running.

After two days and sixty miles through the canyon, the cliffs subside to gently sloping benches, on which grow the first spruce forests we have seen in more than a week. I am glad to see them again, to hear their needles brushing in the wind. I feel now a pang of homesickness I have never known before—not for family, or friends, or foods, but for the airy clatter of sycamore leaves, the rustle of laurel sumac. I have never paid enough attention to these sounds.

Less than an hour after we begin documenting the trees, Russell receives a satellite message from Roman: A large storm is coming. We should get to the airstrip as soon as possible, before the river floods. After coring a few more trees, we jump back into our boats and continue downriver.

Clouds gather in the afternoon, and into the next morning. The sky grows darker and darker. By the evening of July 26, as we pitch tents at the mouth of the Kugururok River, the first sheets of rain have begun to fall.

6

July 29: The storm is here. For two days wind and rain have gusted and spattered our tents; the Noatak, which drains a watershed larger than Maryland, has swallowed the airstrip in its dark, raging torrent. Meanwhile we are running low on food. We expected our resupply to come in with Roman today, but foul weather is predicted through the end of the four-day forecast provided by our satellite device, and it will take days of sun to dry the airstrip. Thankfully, the boats allowed us to pack extra rations for this leg, and this morning we inventory what remains.

"Four packets of grits…"

"A sixth of a jar of peanut butter…"

"Two tablespoons of powdered milk…"

All told I have close to seven thousand calories left. Laid out before me, it looks substantial, but my metabolism—ramped up from a summer of travel—is monstrous. I've been losing weight on roughly 3,500 calories a day, and did not have weight to lose. We decide to budget our remaining food for ten days, which for all of us amounts to roughly one-fifth rations. The sight of my allotment for today—an instant coffee with a tablespoon of milk, a scoop of peanut butter, a four-pack of Oreos, a granola bar, a one-hundred-calorie packet of instant grits, and one bite of my chocolate bar—triggers a hint of panic.

Next we inventory our fuel, which comes out to six or seven days' worth of normal use. If the channel we've been drinking from washes out, we will have to start boiling drinking water, and we'd like the luxury of making hot water in the afternoons to fill our stomachs. For that we need firewood, so I go to the forest and gather the driest boughs I can find, then haul them back to our tents to dry further in the warmth of our body heat. On the way back, I find bear prints—another concern. The bet we make, when sleeping with our food, is that a bear will be too nervous to enter our strange-looking tents. But after several days in one spot, this newness wears off, Roman has told us, and a bear might grow bold.

We divert ourselves by considering how we would find food if stuck here much longer than ten days.

"We're in a pretty diverse area," says Allen. "We're on a floodplain right now. There's forest right over there, and then there's tundra on the other side of it. I'm sure there's food around."

Yet we have only the faintest idea where, or how, to get it. With the river roiling, the fish will be hiding, Julia says. She knows how to snare hares, but only in winter, when their paths are etched in the snow, and we do not have trapping wire anyway. Allen has spotted dandelions along the river bar, but these are few and far between, and we aren't sure that their edible leaves are worth the energy they take to gather. Two days ago we came upon a porcupine, and perhaps, if we grow desperate enough, we could track and club it to death, singe the quills off over a fire, and roast the meat.

It is with awe now that I recall some of the traditional lifeways of

Northwest Alaska's Natives, who survived here even in the days before guns, nylon nets, or freezers. A storyteller named Paul Monroe told a twentieth-century anthropologist of how he and his family survived a sparse winter by tricking hares: They cut a board of wood, blackened it with charcoal, and threw it over the creatures, who—thinking the board was a raptor—would run into their dens, where the hunters could dig them out. Richard Nelson wrote of watching a Koyukon hunter follow a bear's tracks "made in frozen moss, then covered by two feet of undisturbed snow."

The methods varied widely, because animal populations were always fluctuating. Snowshoe hare numbers exploded and crashed with such volatility that the Koyukon said they fell from the sky; muskrats, for the same reason, were said to materialize out of the mud. Some years the salmon and seals failed; other years it was the caribou. To persist here required a depth and breadth of knowledge, and a level of adaptiveness, that is difficult even to imagine.

Our best option now is to conserve energy and watch the weather. So we read, play cards, and make jokes that grow dumber as the hours pass.

*　　*　　*

July 31: It is our fifth day tent-bound, and our third on one-fifth rations. For dinner last night, I ate half a packet of instant ramen with lots of hot water and then tried to sleep, but all night long my stomach rumbled. I dreamed of food and woke up drooling. More than hunger, it is the uncertainty that nags my thoughts. Our latest weather forecast says that the rain may peter out today, but that another large storm will arrive the day after tomorrow. There is no end in sight, and we may soon have to consider stretching our rations further.

Along the forest's edge I try digging up Eskimo potatoes, but there

NORTH TO THE FUTURE

is almost no meat on them. Julia noticed rabbit droppings in the forest yesterday, and now, mostly to pass the time, she is carving a beautiful bow and arrow from willow. (We have the shotgun, but even if we thought it wise to use our five rounds to hunt, the slugs would burger-ize anything smaller than a caribou.) I think now and then of the porcupine. As much as its meat, I want the knowledge it's acquired of the berry patches and mushrooms and root beds—a knowledge that was legendary among locals. "Porcupines know all the country," a Koyukon man told Richard Nelson.

When we break our fast this afternoon, Allen teaches us a Yup'ik word: *boguk*. "It means to chew every last bit of meat and cartilage off of a bone." I *boguk* my last bag of chips, tearing the plastic open and licking the aluminum lining for every trace of salt and grease. Toward evening, the rain pauses, and we wander into the woods in search of berries.

In less than a mile we come to a thicket of low-bush blueberries, which leads us to another thicket. Soon we are crawling through the sopping moss, stripping tiny, half-ripe berries by the dozen, chasing the next handful of gleaming fruit. After a trancelike stretch—two hours, Russell says—our quarter-gallon food containers are full and our lips are stained purple, and we decide we should return to Maddy, who is watching the tents and may be getting worried.

Rising, I grow lightheaded and find that I have lost all sense of direction; I follow Julia through the forest like a toddler. Twenty minutes later we are back in the warmth of our sleeping bags, still thinking of food. The berries have done little to take the edge off our hunger; what our bodies crave is protein, carbs, salt.

I realize now that I have never truly been hungry before. And though my body aches and the world spins when I stand, this feeling is a pale shadow of the hardships endured by this land's earliest inhabitants. Some years the down cycles of many animals—the caribou, the hares, the salmon—coincided, and then no amount of knowledge or skill was enough. A little over a century ago, an Iñupiaq man came upon his wife a few miles from here, leaning against a tree. Above her hung half of a human body. She had spent the winter upriver with her family, during the same caribou decline that decimated the Nunamiut farther east, and the rest of her family had starved to death or been killed by her brother for food. The hanging body, which she'd been eating, was one of her parents.

"You go home," she told her husband. "I am not fit for that anymore."

Cannibalism was rare here; more common are stories of sacrifice and kinship. Sometimes women even breastfed their husbands to give them the strength needed for hunting. But food shortages were a fact of life and often meant boiling the grease out of bones, searching the bases of trees for frozen birds, or even eating willow bark or animal droppings. Often the best strategy was to flee to neighboring country while one still had the wherewithal to do so.

Our predicament is mild in comparison, but it occurs to us now that fleeing may be our best option, too. Scanning the map, Allen notices that the Native village of Noatak, the only village along the entire length of this river, is just over fifty miles away. There we would surely be able to buy food, and at the rate the river is flowing, we could get there in a single day.

The risks are clear. The river is clogged with trees and sharp sticks that could pop our fragile boats in an instant, and if that were to happen, we might be swept under. Given how little we've been eating, the risk of hypothermia has also multiplied. "Things could get weird pretty quick," observes Russell. But as soon as the water begins to drop, we reason, the river will stop sweeping up so much debris, and the risk of popping a boat will drop, too.

NORTH TO THE FUTURE

When we arrived here five days ago, we planted sticks along the river's edge, and we've been monitoring them every six hours to check the water level. For the last several days the river has consistently risen, but the rain has stopped this evening, and according to the barometer on Russell's watch, the air pressure is climbing. We agree to wake up at five thirty a.m. to check the sticks again: If the water has gone down, then we will make a run for it.

* * *

Wind batters the tent all night, but does not bring rain. Sometime around four a.m. I drift to sleep. At five fifteen I wake to the sound of Russell unzipping his tent, heading out to check the sticks.

Fifteen minutes later he returns with promising news: The river rose in the night, but he can tell by the high-water mark that it has since peaked and begun to drop. Meanwhile the atmospheric pressure is still rising. With another storm arriving tomorrow, this seems our only chance at escape for the foreseeable future.

"How's everyone feeling?" asks Russell.

There is a moment of silence. Julia and I look to each other; then we all speak up in agreement: *We should go.*

Russell uses the satellite phone to call Roman, who approves our plan. We begin to pack. Before stowing food, Julia suggests that we eat a full breakfast to fortify ourselves for whatever lies ahead. No one objects. For myself, I boil a packet of instant grits and add muesli, a handful of mixed nuts, and three generous scoops of powdered milk; cereal has never tasted so rich. Half-satiated for the first time in four days, we begin to dance and sing. I pause to snap a few photos of the others, spinning and cartwheeling over the muddy floodplain, their hair tumbling, eyes wild.

But the Noatak, when we reach it, sobers us. Clear and placid six days ago, it has transformed into a ravenous deluge, ripping through the forest and spitting up wood. We watch a full-grown spruce tree careen by, its green boughs thrashing.

"This is a risky move," says Russell, eyeing the tree. I sense that he is feeling the weight of the decision, thinking it over again. "We gotta stick together," he finally adds.

Hugging each other, we fasten our life preservers and nose our boats into the current.

It grabs us in an instant, sucking our rafts into a train of long, slow waves. We roll up and down like bits of loose flotsam. *"It feels like we're on the ocean!"* shouts Maddy. But we do not dare stray from the main current. Roman warned Russell this morning to avoid the sides of the river, and it is now obvious why: There are none. Water has washed over the islands and banks, turning the entire floodplain into a roiling sheet of gray. Rafts of logs swirl in ghostly eddies; bluffs cave and crumble, bringing mats of moss and trees down with them. Patches of still-rooted shrubs shudder in the current, and no doubt shorter ones lurk beneath the surface, their tips more than sharp enough to sink us. After a few miles I notice an unnerving hissing sound; Russell and Maddy look down at their boat, too. "I think it's just silt," says Julia. Stirred up sediment grating against the raft. We have heard this sound in cloudy rivers before, though never quite so loud.

As we gain our bearings, I find myself grinning. The sky is dark, the water cold and menacing, but the land has brought something out of us, exposed some reservoir of strength and skill we did not know we possessed. Soon we are hooting and howling into the mountains. Gradually the dream of escape grows more and more real, until it occurs to me that Noatak Village may offer something more than chips and chocolate: A window into one of the oldest cultures in the Americas, about which I have read—but only read—for years.

7

Late afternoon, after fifty miles on the swollen river, the first rooftops come into view, a smattering of blue and red and gray poking above the trees. Beyond them lies a modest harbor, where small open boats with outboard motors are moored to a bank of concrete-filled bags. A sign says NO VISITORS. There is no one to be seen.

The villagers of Noatak are descendants of two Iñupiaq nations: the Nuataagmiut, who spent most of the year above the Noatak Canyon, and the Napaaqtugmiut, "the people of the spruce," who lived mostly southwest of the canyon—harpooning seals on the coast, netting salmon and whitefish on the lower Noatak, and hunting caribou upriver. At the end of the nineteenth century, both nations suffered from the same food shortage that afflicted all of Northwest Alaska, and most either fled or died of starvation. When Quaker missionaries boated upriver and built a school at one of their summer camps in 1908, many of the survivors gravitated there, and within two years nearly the entirety of both nations had peacefully settled. What ensued was a period of stark material and cultural transition that continues to the present day.

The NO VISITORS sign gives me pause. During my first summer in Alaska, I spent a week in the predominantly Iñupiaq town of Utqiagvik and found myself wondering, each night, why I had come. Most locals were polite, but it became clear that open-ended curiosity was not a legitimate reason to ask for their time. They had seen more than enough white men come north with big dreams and little ken.

In the nineteenth century, American whalers brought rifles and steel to Northern Alaska, but they also brought disease and alcohol, and decimated the whale and walrus populations. Missionaries came with books, medicine, and a reliable source of outside food, but they also denigrated Native lifeways and forced children to speak English, cutting a generation off from its cultural roots. And while the US government's policy toward Natives has been progressive in Alaska relative to its shameful legacy in the Lower 48, the effects are complex. In 1971, thanks to the lobbying work of Willie Hensley—an Iñupiaq man who taught himself the law in order to prevent his people from being written out of it—Congress passed the Alaska Native Claims Settlement Act, allocating 44 million acres and nearly $1 billion to thirteen brand-new Native corporations, organized to serve the interests of the state's roughly eighty thousand indigenous people. This ensured that they held capital and an economic stake in Alaska's resource development, but it also required them to adopt a concept of private ownership that was antithetical to their nomadic practices. One Native leader compared this to being born adult, and then having to learn to walk and talk.

Wary now of imposing, we consider continuing downriver, but our stomachs win out. Leaving Maddy and Julia with the boats, Allen, Russell, and I begin up the bank, following the only road into town. As we round a corner, several children run up to us; bashful but friendly, they eye us too long in the way young children do. Following their outstretched fingers, we soon come across a middle-aged man with a ready smile; he tells us we are welcome to camp by the airstrip. Returning to Maddy and Julia, we pack up our boats and head into town.

In under a minute we reach the heart of the six-hundred-person village: a post office, a Quaker church, a school, a general store. All but the church are closed for Sunday; nearly seventy miles above the Arctic Circle, we are back in the world of Sundays. Though the furthest houses are only half a mile apart, villagers zoom by on 4×4s. A hunter stops to ask if we have seen moose upriver; others slow down to say "Welcome to Noatak!"

Caribou racks hang over the open front doors of houses, which are built on stilts to avoid thawing the permafrost. Dogs pull at chains. TV dishes cling to colorful metal roofs, and rusting vehicles—too expensive to ship out—clutter front yards that tail off into stands of willow and muskeg. A quarter mile from the harbor, the houses end, and we come to a row of fifteen truck-sized gas tanks, each newer than the last. Here a driver points us down a dirt track, leading to a grassy terrace above the river where we can camp. "Welcome to Noatak!" he adds.

The adrenaline fast fading, we pitch tents and devour our remaining food scraps. Then I drift off to the sound of the breeze on the tent, and the pattering of the rain, and the whines of dogs quavering through the valley.

* * *

In the morning we head back into town for food. The sky is overcast; dogs lap at day-old puddles. An obese woman motors by wearing AirPods, and a white man—one of only a few among Noatak's 95 percent Native population—drives alongside his leashed dog, almost clotheslining me. Near the store, a passing man invites us to get water at the pump house across the road, and a few minutes later I decide to take him up on the offer. Following his directions to a nondescript warehouse, I find the door opening just as I reach it; the man is standing there, ushering me toward a pot of hot coffee.

His name is Paul Walton, and he's managed Noatak's water and sewage system ever since he rebuilt it three decades ago. Pacing around the room, he turns back to me from time to time with a smile so wide, so electric, that I wonder how much energy it takes.

"This is holy water," he says, gesturing down at my cup. "More than ten times cleaner than the state requires." He continues to pace. "Forty, fifty below, we're pumping water, twenty-four hours a day, three hundred sixty-five days a year. It's a hard job. Gotta be chasing your tail all day long. But somebody's gotta do it."

As I sip coffee, he gives me a quick tour of the operation—an ornate matrix of pipes and tanks and shining gauges. Three hundred miles removed from the nearest highway, in one of the harshest environments imaginable, the logistics of sustaining a modern standard of living are soberingly involved. Paul says he will likely need to rebuild much of the system next year because, as rainfall increases, the river erodes more and more of the bank on which Noatak Village sits.

"Last year I had a hundred feet from the river to my transmission line," he says. "Now I just got about fifty feet. One corner probably will be exposed again soon, and I'm gonna have to try to reengineer a new line to my wells. So that's a burden."

But Paul is old enough to put such a burden in perspective. He was raised by grandparents who were born in 1897 and 1901, when the Napaaqtugmiut were still nomadic. Prior to settling here and investing in infrastructure, there was no water treatment plant, and no one expected things to last more than a few seasons; it was customary to build a new dwelling every several years. And while river erosion has accelerated rapidly in recent years, it has always shaped life in Northwest Alaska. The anthropologist Edwin S. Hall Jr., who spent a year in Noatak from 1964 to 1965, reported that the river had already forced the village to move more than three hundred feet since its founding.

"Long ago, we didn't own land," Paul tells me. "We just use it until we go, and we give ourselves back to it. Then the white man came and we had to get a little smart. Too much politics now, I guess." He turns bright again, flashing his default smile. "But it's a good country. Good life."

* * *

I thank Paul and join the others at the general store, a windowless warehouse run entirely by Alaska Natives. The shelves here are stocked with everything from peppers to rice to kiwis, as well as candy, soda, and tobacco. But while it was possible, until around the 1990s, to import these goods via barge, the river has since grown too wide and shallow for large boats, and now everything must be flown in. Prices reflect this: A one-pound jar of Jif Extra Crunchy peanut butter costs $5.65. A quarter-pound of salted butter is $4.85. A gallon of gas—priced around $4 in California when I left—is $8.99.

The cashier defrays our expense with an employee discount. "It rained twice this summer," he remarks, after we explain why we have come. "Once for forty-seven days, and once for thirty-two."

"People are asking, 'Where's the summer?'" says Ben Arnold, sixty-three, whom I meet outside the store. "'Where's the sunshine?' We were gonna take children downriver tomorrow to get salmon, teach them how to cut it up, how to dry it. It was supposed to be a three-day program, showing families the cultural skills. But we had to cancel because there's no eddies for catching salmon. The water is too high."

More generally, he says that animal migration patterns—knowledge of which lies at the core of the culture—have grown unpredictable.

"*When* and *where* you get the cultural foods—that's the important thing," he explains. "Right now, today, it's a guessing game. You never know when it's gonna happen. Otherwise we're dependent on *that*," he says, turning to the store. "And it's expensive. A lot of people are fortunate to get assistance from food stamps, but they don't have jobs. Or they depend on the permanent fund, or other programs. And it seems to be going toward that direction." He turns back to the valley. "Depending on what's going to happen out here."

*   *   *

We spend two days in Noatak Village, during which the weather gradually clears. A few preteen girls catch wind of us and parade us around town, snapping selfies on our phones with Cheeto-covered fingers. They are sweet and clever, with shining eyes and ready laughter, and I wonder what this place will look like when they are Roman's age, or mine.

I step away often to interview those who care to speak. People talk of late winters, foreign birds, and berries the size of grapes; mudslides, unseasonable storms, and treacherous travel. Everyone has grievances—from the banal ("All that wind messed up my satellite dish, and now we can't get any channels, hardly!"), to the aesthetic ("Usually we'd have an awesome sight to see come breakup—thick chunks of ice flowing down the river. This year there was just a lot of dead wood!"), to the deadly ("Our winter trails are getting thinner, rotting away faster. There's a few people that we have lost these past couple of years due to the ice conditions.").[1]

---

1 Two winters ago, a couple and their eleven-year-old niece broke through the ice and died while snowmobiling back from Kotzebue. Some villagers blame the deaths on unusual ice conditions, others on the driver's inebriation.

NORTH TO THE FUTURE

Some seem, on the whole, unperturbed. "As long as our freezers are full, we're okay," says one man. "This year we've just been getting a little more wet, though." Others echo Ben Arnold's concerns—that the land's transformation is threatening a way of life already under intense pressure.

"We go by seasons, the *whooole* year around," says Thurston Booth, one of Noatak's most respected hunters. Sitting on his ATV, he traces a circle in the air with a smoldering Winston cigarette. "Season after season. Our parents been doing it, and our grandparents been doing it, and they show us how to do it. I do it every year, ever since I get to learn.

"We used to get ready for the season to start. And we *know* when it starts. Nowadays, we don't. About fifteen years ago, that's when we start to see a lot of changes. We see the riverbanks are falling in big chunks. Ice out there don't last long anymore. Our trapping season's only two, three months now."

He looks down at his toddler grandson, who is playing in the mud. "Nowadays there is so much microwavable food. Easy for them to, you know, just throw it in the microwave and heat it up. It's so easy, and it's getting them trapped."

Elders also bring up smartphones, which are as prevalent here as in LA.

"We used to be out everywhere before these phones came," says Leonard Vestal, father of six. "Nowadays the kids have been staying home. But it helps! You need help, got a problem? Now we can contact everybody. These new cellphone batteries, they last a long time." He adds that he and his children have been getting into PUBG, an online shooting game. "That one's kinda fun."

One elder suggests I talk to Lonnie Arnold, Ben Arnold's nephew, who at thirty-one is already a pillar of the community. It turns out that he is the cashier who gave me a discount at the general store. He also hunts year-round, volunteers on the regional search and rescue team, and works as a gravedigger.

"Our climate, our environment, it sure has changed a lot," he says, when I run into him along the main road on our second evening in town. A tall, slender man with a buzzcut, he speaks so slowly, so earnestly, that for a moment I wonder if something terrible has just happened.

"We try to adapt. That's what we've been doing for many generations, is trying to adapt. Adapt with our lifestyle change, adapt with our seasons, adapt with the environment. What's going on? What's it teaching us today?"

From a certain vantage, it is easy enough to look at the derelict vehicles littering the yards, the unemployed adults wandering town, and the kids staring down at glowing screens and recognize, in Noatak Village, the clash of two civilizations in decline. For decades, Noatak, like virtually all of rural Alaska, has depended heavily upon oil—both logistically, for importing goods and motoring to hunting grounds, and economically: to pay for social services and to build things like the new $28 million public school, which glitters on the far side of town, just past the raven-swarmed garbage dump. But talking to these people, one finds that there survives here a perspective that has become rare today: a view of nature as something other than a collection of resources, or a timeless garden of Eden, or any other static backdrop to the human project. A recognition of this planet as the substrate of life itself: a thing both sacred and profane, and just as wild as ever.

There along the main road, on our last night in town, Lonnie tells me a story that I know will stay with me for years to come. He holds each word in his mouth, breathing through them like prayer.

"This was back in 2012," he begins. "I was traveling back home from the next village over the mountains, Kivalina, and it was wintertime,

February. It get stormy on me, and my [snowmachine] broke down. I had no tent. I had no food. I had no water. Nothing. Just my backpack with my extra clothes. It was only supposed to take me an hour and a half to come home. And I was out there four and a half days. Close to twenty-below weather, hundred-mile-an-hour winds. Real fine snow like sand, so I could barely see my hands. I don't know how I survived.

"I tried walking back in that storm. The snow was waist-deep, draining all my energy. I try to keep going. I just try to keep going. Try to keep walking and finally I couldn't no more. And I come to this one big knoll, one hill I come to, and I stop right at the bottom of it. I was so tired, couldn't do nothing.

"Finally I just start praying real hard, and then I don't know why, I just felt a boost of energy to get up and make me a little…my grave, I say. My grave I made me. And I got in there and say, 'Lord, if they're gonna find me, they're gonna find me right here, dead or alive.'

"And as the minutes passed, as the hours passed, and as the days passed, when that storm was around me it covered me in snow. There were animals around. I could hear wolves. Four and a half days I was out there. By the time they found me, my legs were all frostbit. They thought they were gonna have to amputate.

"Those nights were so hard on me. I wasn't expecting to be found. I thought I was gonna go out there…" His voice cracks and he begins to tear up. "I gotta let it out. Sometimes I gotta keep talking about it so it don't hurt me no more. It always come back to my head from time to time.

"I was floating in bright, bright clouds. And all I heard was 'It's not your time to go, son. I got stuff for you to do. I gonna send you back down, son.' I sure didn't want to come back, but then he sent me back down. Because he say I got stuff to do for him."

8

August 3: In the wake of the storms, a hot stillness is settling. Steam rises over the river; bugs whine. Allen flew out yesterday, and this afternoon the rest of us gather at the airstrip, where several medical workers, in town to test for COVID, are awaiting a flight out. When Russell asks if the new Delta variant has reached the region, one of them exhales smoke from the side of her mouth: "*Oh* yeah," she says. Later an earnest young villager comes by to warn us against vaccination. "The last days are coming," he says, "and those who get a computer chip implanted the Lord will not forgive."

Just before five p.m., our pilot arrives to fly us back into the field. WELCOME TO NOATAK, a sign at the airstrip tells us for the last time; we board the plane and whiz over it, watching the cluster of colorful metal roofs shrink to nothing. Our terse pilot—who has gotten little rest since the rains cleared—retraces our route up the Noatak, and then tacks west, up a river the villagers call Kuugruuraq, but which cartographers named (and misspelled) "Kelly," after a nineteenth-century prospector named Kelley. After a dozen more miles, the pilot zeroes in on a willow-covered island, and we land on a strip of gravel barely wider than the plane's chassis.

Three men are waiting in the willows. One is Ray Koleser, who worked for the US Forest Service for three decades, and is one of only a few outdoorsmen Roman's age who can keep up with him. (The two hit it off when Roman learned that Ray's email address began with *Picea*, the Latin genus for spruce.) Next to him is Jason Geck, a young professor of environmental science at APU. And in front stands Roman: Out of solidarity with us, he has kept his tumble of white hair and his bushy, snowy beard, and as we emerge from the plane, he wraps us all in gruff hugs.

* * *

For three days we camp by the airstrip and do fieldwork nearby—sawing shrubs, counting seedlings, measuring and coring mature spruce trees. The forests are flourishing here, as they seem to be everywhere in the west. So are the bugs. The mosquitos have thinned, but a welter of biting gnats and flies have replaced them, and in the stillness they form a thick haze, finding their way between the buttons of my shirt, leaving welts up and down my torso, burrowing into my beard and scalp. One day I see even Roman grow agitated and dart away from a swarm. Despite the heat, we stay covered, sweating and bleeding as we work. Lowering our pants to relieve ourselves is a kind of torture.

The berries are flourishing, too. Like so much life here, they cycle wildly from year to year, and this season—watered by the rains, ripened now by sun—they grow to Brobdingnagian proportions. We pick blueberries the size of coins while we work, popping them into our mouths and saving handfuls for morning cereal. The bearberries are huge, too, and so are the cranberries and the crowberries—a black, bland fruit that the Koyukon called *deenaa tt'aas*, "scattered charcoal."

Best of all are the cloudberries, a salmon-colored fruit that grows low to the ground and tastes creamy and sweet, almost like yogurt.

Since the food shortage, even normal rations leave my stomach growling; I gather mushrooms, too. The Natives of Northwest Alaska did not eat mushrooms, probably because some that grow here are poisonous. All were tabooed by the local Iñupiaq shamans, who called them *argaiġnaq*—"that which causes your hands to fall off." But Roman has taught me to identify edible boletes, the caps of which look like golden-brown pancakes, and I pick them as I go. Later I sauté them in powdered butter, salt, and pepper.

After fieldwork each day, we bathe in an icy channel of the river, stirring up sand around huge salmon. Julia and I bought several lures from the Noatak general store, and now we try them, but these fish are too heavy for our line; we lose all three hooks in an hour. For our efforts we earn hundreds of bug bites. My body is on fire.

But there is something nourishing in the mere sight of these fish, alive in clean water. Something moving in the profusion of wild fruit. It seems strange to me now that most of the world's major religions locate the divine in remote deities, and mediate this divinity through symbolic abstractions: the cross, the Star of David, the Khatim. Holiness, in these cosmologies, is a far-off thing, all but inaccessible to the senses; earthliness is often synonymous with spiritual corruption and barbarism.

Many traditional North American cosmologies, by contrast, are rooted in this world—ground, water, air. For the traditional Koyukon, all plants and animals have spirits, but the most powerful one, *sinh taala'*, emanates from the earth. For the traditional Navajo, another Athabaskan nation whose progenitors must have once passed through here, the wind came first, before earth and humans and deities. It marks human flesh in the whorls on our fingertips and toes, and the cowlicks on our heads. Breathing is thus a holy act, an exchange with the essential life force from which we came, and to which we will return. (The Navajo are not alone in this belief: The ancient Greek word *psychê*, the Latin word *spiritus*, and the Hebrew word for spirit—*ruach*—all also mean "wind.")

I have realized, over the past three years, that I have no desire to turn back the clock—to live under the spell of shamans, or to huddle through an Arctic winter in a skin-and-sod dwelling. When asked in the 1970s if his people weren't better off in the days before modern technology, Chief Henry—leader of the Koyukon at the time—was reported to have replied simply, "Did you ever have to keep alive by eating ptarmigan droppings?" I have experienced, moreover, the power of modern tools to open our minds to the land: Matt's maps and Logan's remote sensing studies have revealed processes we can't directly perceive. Roman's iPhones have allowed us to gather vast amounts of useful data, and the digital library downloaded on my Kindle has provided important context, priming me to look for clues. When I return home, I will spend long hours indoors, staring at a screen, trying to turn these notes into a story worth sharing. Sometimes taking a step back from the world helps to see it more clearly.

But sitting here now, watching the light on the mountains, I feel no need to take a step back. No need for otherworldly abstractions to carry the heaviness of being, or to find a reason to strive for goodness. The sacred lies all around—in the rushing of the river, and in the upturned willow leaves blowing in the wind.

* * *

August 7: Morning comes cold and clear. We break our camp of three nights and set out on foot, following the Kelly River north into blue mountains. At a fork in the river we turn west, up a tributary that our maps call Wrench Creek, but which the local Inupiat call Kataq—"fallow." Today it is anything but: The tundra sparkles with fruit, ambered by the night frost. A grizzly on the far bank is too busy munching to acknowledge us.

Overnight, the tundra has begun to turn. Yesterday all was green; today the willow leaves are yellow and the dwarf birch is speckled with red. Stalks of sourdock shoot maroon from the soil, and the bearberry leaves have begun to smolder. The chill of fall is in the air.

The sun is low when we cross a fork in the river and climb to a naked ridge, threaded by caribou trails, covered only in a thin scrub of dying wildflowers. As we set down our packs on a promontory, Julia notices a nubbin of shiny rock. She bends down to inspect it just as Roman calls us over: At his feet, surrounded by stone flakes, lies a gleaming arrowhead.

Fanning out to comb the ridge, we find several more glass-like slivers of chiseled chert. Hunters once camped on lookouts like this one, carving arrowhead and spear tips as they surveyed the valleys below for caribou and sheep—and, before that, bison and perhaps even mastodon. It is generally impossible to date such "surface scatters," as archaeologists call them; these points could have been made by the first people to set foot in the Americas, or by nineteenth-century Iñupiaq hunters. But whether old or new, they are today out of place. Hunters would have camped far above the caribou thoroughfares, waiting to ambush from on high, but now a caribou highway runs over the artifacts. Roman surmises that the shrubs that now choke the valley are new, forcing caribou to travel up along the valley's rim. Caribou, after all, do not like shrubs and trees; branches catch their large antlers, miring them down and limiting their range.

For this reason, Roman suspects that caribou will be losers of the boreal forest's advance, as will those who hunt them. There is some limited evidence for this: For at least ten thousand years, humans have gathered at Onion Portage to hunt caribou, but the area has grown shrubbier, and for several years now the caribou have not come. Scientists have put radio collars on the animals, finding that they've been traveling up valleys that remain relatively barren. The Western Arctic Herd, the largest in Alaska, has shrunk by two-thirds since the turn of the century, and possible explanations include disease, starvation, and slowed migration, all of which are associated with climatic swings.

But we don't really know how caribou will respond to the new climate. Their populations have always swung wildly: a herd of five hundred thousand can drop to one hundred thousand one decade and rebound the next. "There is no one historical moment when the herds are not either recovering or preparing to falter," writes the Arctic historian Bathsheba Demuth. "I wish I had a better answer for you," said one expert, when I asked the reason behind this volatility. "If you asked twenty caribou biologists, you'd probably get twenty different answers."

We camp on the ridge, taking in the hunter's view: a 270-degree panorama of both forks of the river, the lower valley, and a red mountain below that smolders in the sunset. Over dinner, Julia and I talk about the caribou.

"Growing up, I had this feeling that there was nothing else to explore, because we had all this information and data," says Julia. "It seemed like people had an answer for everything. But this project has shown me how much we don't know. And it's a *relief*. I'm so relieved to learn there's so much unknown left in the world."

In the night I step outside the tent to pee, and find the red mountain frosted in blue shadow. A stillness has settled, and for a few breaths I am submerged in a silence millennia deep.

* * *

August 8: Today we climb northwest into the mountains, finding more spear tips and arrowheads, but few caribou. More common are the moose, who stick out above the willows like drifting islands of fur. Moose populations are not quite so volatile as caribou, but nor are they dependable. The species disappeared entirely from Northwest Alaska in the nineteenth century, only to reappear in the 1940s; by the 1970s they had become the single most important animal resource for many inland hunters, more so even than caribou and salmon. It is still unclear why the moose left and returned, but many scientists expect that they will thrive in the new Arctic, because they subsist on woody plants that stick out above the snow in winter. What range the caribou lose to shrubs, the moose might well gain.

We camp along the creek, beneath an arrow-strewn bluff. The zipper on Russell and Maddy's tent, sticky for weeks, has finally broken, and they sew it up with dental floss—a fix that keeps the rain out and the warmth in, but requires them to roll through the bottom. In general our outfit is wearing thin: our shoes and clothing are full of holes, and we all are managing injuries.

In the morning we follow a glimmering creek into highlands, a bare realm of tundra and rock and wind. On the slope above us, an absent-minded caribou nearly bumps into two camouflaged grizzly cubs; for a moment we brace for attack, but without their mother the cubs freeze, and the caribou darts away.

Higher we see more caribou. They have started their southward migration, filtering through the passes by the hundreds, "like islands of smoke," as the Alaskan poet John Haines wrote. Most travel in small groups of a dozen or fewer, shuttling young calves that are all fur and bone; others travel alone. A few are so tortured by insects that they run within an arm's length of us, trying to offload pests. We peer into the ragged holes in their backs, where warble flies have mined them for energy.

This evening, as we try to sleep, Julia and I start to the sound of a heavy creature charging toward us. Lurching upright, I grab my bear spray and unzip the tent just in time to see a lone caribou huffing past. Bucking wildly, shaking his head, he runs and runs into the twilight.

* * *

August 10: The sun's path has shifted decisively now. At night, the darkness of winter creeps over the land, sending films of ice across the puddles, freezing our socks and shoes. Clouds of ice mist tinkle the stiffened fabric of our tent. During the day, amber light warms my skin, but the wind blows and the shadows encroach. I sweat and shiver at the same time.

We continue west, crossing slopes of talus that grade into windswept mats of mountain avens, purple rhododendron, and dying forget-me-nots. Down below slides a black-silver creek. Along its course lie fossils of seashells, perhaps 80 million years old, and spare groves of monkshood—the flowers of which look like purple butterflies, and are deadly poisonous. From branch to willow branch flit golden crown sparrows: "*I'm / so / lone- / ly,*" coos Julia, imitating their song.

In a gully we find a two-month-old caribou calf, ripped open and steaming into the tundra. Atop the bloody mess sits an enormous golden eagle, tearing at the fresh organs. Roman can tell—from the integrity of the bones, and the cuts in the stomach—that this bird, and not a wolf or bear, has made the kill. We can only guess at how a twelve-pound raptor took down a sixty-pound mammal, but it must have been painfully slow.

In the afternoon, a grizzly appears while we are sawing down willows, and we sneak away. A few hundred yards up the creek another figure catches my eye: a soot-colored wolf. Then a second, creamy white. Seeing us, the pair lopes up the mountain, looking back now and then with glowing eyes.

I am not on pixel-walking duty today, and am free to walk for hours without recording anything. The land ripples my senses, articulating new colors, textures, feelings. The wind blows at embers banked somewhere deep in my mind. I came to Alaska, I have lately realized, because there was a hole near the center of my life, a deep uncertainty about my place in the world. I did not believe in anything; a lot of days passed like shadows. I now find, in these clearest moments, that a definite pattern has taken hold of my thoughts, and it is something like faith: not in an all-powerful savior, or the unimpeachable perfection of nature, but a faith in the existence of a world beyond our species. A life that will endure long after we are gone.

* * *

In the evening, after we have pitched our tents, a grizzly lumbers into the valley—the biggest, Maddy says, that she has ever seen. Great shoulders rolling like a silverback, it scatters the caribou. Leaving behind pack and gun, wearing only flip-flops, Roman wanders out after him.

I take a canister of bear spray and follow at a distance, watching Roman watch the bear, wondering what he has read in the animal's gait that tells him it will not charge. He has taught me a lot these past three years. "You look like a different person," he said a few days ago, and from his tone, I knew he referred to more than the gingery beard that has overgrown my face or the bug scabs and sun freckles. *Eagle eyes*, he has called me, and when we encountered the two grizzly cubs yesterday, he asked me to keep my eyes out for their mama. "You're good at finding them," he said.

But as much as I have learned from Roman, most of his knowledge remains his alone, somewhere beyond the reach of words. Time and again, I have staked my life on his intuition with something like blind trust. Now I can only watch in awe as he stands at the valley's rim, looking down at the great bear, reading signs I have not yet learned to see.

9

The next day, we descend out of the highland to the headwaters of the Wulik River, where our pilot has deposited our final barrels containing food and packrafts. The following morning—August 12—we inflate our rafts and begin the final eighty-mile float to the ocean. The river here is low and fast, and flows in braided channels that in places are less than a foot deep and only a few feet wide. In canoes it would not be possible to float, and in packrafts it is only just. We flit over stones, chasing narrow tongues of bubbling water, whizzing past banks and overhanging willows.

Creeks pour in from the mountains; the river deepens. By afternoon the gin-clear flow has grown to an emerald torrent, and the valley has narrowed to a gorge. Soon the water begins to boil and foam over boulders, drilling down into dark hydraulics, and I find myself white-knuckling through the biggest rapids I have ever floated. Realizing that a fall here without helmets could turn deadly, Roman pulls over into an eddy, and we begin carrying our rafts over limestone slabs. But after about an hour of treacherous wrangling, the sides of the canyon turn to sheer cliffs, and we are forced back down into the water. *Look where you want to go,* echo Roman's words in my head; I keep my eyes pinned on his splashing boat.

And then, suddenly, we are through, and the mountains subside, and the horizon stretches out before us, a long, rumpled apron leading to the sea.

*   *   *

August 14: Nearing the coast, we see groups of sandhill cranes unfolding their spindly frames, releasing clattering calls. They come here each spring to bear offspring in the endless summer day, and are now beginning their fall migration south to Texas, Florida, Mexico. While nesting, their long, sharp beaks become weapons, but the Koyukon say that if you get close enough to one, and emulate their song, they will dance for you.

The cranes are a small part of the unique profusion of life that gathers along this coast every summer: some 20 million migratory birds and a disproportionate share of the world's seals, walruses, and whales. More than 40 percent of the US's fish and shellfish catch comes from the eastern Bering Sea, which is one of the most productive marine ecosystems on Earth. This productivity is driven in part by the shallow continental shelf, which dredges up cold, nutrient-rich water, and in part by the fast-receding sea ice, on which grows the algae that fuel phytoplankton blooms—the base of the entire food web.

From the boat I see bear prints going back and forth across the beach: There must be salmon. Soon we begin to see them—a few at first, then thousands, all nearing a natural, if excruciating, death. After spawning, their glands release a mortal flood of steroids. (Remove the adrenal, scientists have found, and they can survive for another year.) They lose their luster; their jaws hook into a caricature of themselves; mold grows, ulcers develop, and they begin to decompose alive, revealing fatless white flesh beneath. Around us, many already lie rotting on the shores; others have entered a zombielike state and thrash at the surface or float torpidly below. Within weeks their bodies will fertilize the plants along the shore, feed insects and other small creatures, and in this way nourish their own offspring, as the latter make their way downriver to the sea.

It remains an open question how the salmon will respond to the new climate. Like the caribou, they have always cycled wildly, but along the Yukon River—where numbers have been most thoroughly documented—their population is now the lowest since records began in the 1970s. The state is beginning to prohibit many villages from fishing at all, though it is a staple of traditional life here. Many locals blame the pollock fisheries that trawl in the Bering Sea, but most scientists believe something deeper is afoot. Warming water temperatures are thought to increase metabolic rates and decrease food availability. Recent seabird and whale die-offs hint at tectonic shifts in the Bering Sea's food chain. And all of this heat might be triggering the spread of *Ichthyophonus*, a genus of parasite now found in almost half of Yukon salmon.

We pass more bears; a golden-red wolf; and two groups of fishermen, here for the Dolly Varden trout, which can grow to twenty-five pounds in this river. "If you catch one o' those, you gon' be on a Nantucket sleighride!" says one angler, a large, friendly policeman on vacation from "Katrina country." We begin to hear helicopters, shuttling back and forth from the Red Dog Mine, the largest zinc mine in the world. The feeling of wilderness dissipates slowly, then all at once.

*   *   *

On the afternoon of August 15, our final day in the field, we see power lines. They follow a gravel road that begins about fifteen miles to the west, at the village of Kivalina—a nineteenth-century Russian naval officer's mishearing of Kivalliñiq, the name given to the region by the resident Iñupiaq nation called the Kivalliñigmiut. In those days, Kivalina was a summer camp. Families built moss and willow dwellings on the shore and hunted bowhead whales, belugas, bearded seals, walrus. In 1905, Quaker missionaries—funded by the federal government—built a school there, and within a few years the entire nation had settled year-round. "We got stuck," recalled a local woman interviewed in the 1980s.

Kivalina's inhabitants have, in recent years, become unstuck. The village sits on a barrier island, a two-hundred-yard-wide strip of land between the ocean and a lagoon. It was always vulnerable to erosion, was never a logical place for a permanent settlement, and now, as the sea ice recedes and the ocean warms, the same storms that blow seeds north through the Brooks Range are eating away at the thawing coast. In October 2004, a single storm took forty feet of land the village did not have to lose; the school flooded, and the principal's trailer dangled over the surging ocean. The village had no choice but to begin the long process of relocation.

First the gravel road was built, stretching over the lagoon and another seven miles inland to the foothills of the Brooks Range. Then a school, and soon they will build a power plant and houses. The total cost, paid for by state and federal agencies, has been estimated as high as $400 million, about $1 million per resident. In the meantime, Roman says, the houses are stocked with life preservers.

* * *

Near Kivalina, during a water break, I find myself alone with Roman on the riverbank. Turning to me, he asks a question: Having spent all of these months in the fastest-warming corner of the planet, catching glimpses of dramatic changes and prehistoric continuities, am I now more or less optimistic about the future?

I garble something unintelligible about loss, hope, and the value of wilderness. I am a little ashamed; having asked Roman hundreds of questions over the years, he has finally asked me one, and my answer is hackneyed garbage. But as we continue downriver, I give more thought to the question and find that it touches many, many others.

How fast will the trees move north? What about the shrubs? How much of these soils will thaw, and how much carbon will they emit? How quickly will the glaciers melt and the seas rise? How many of the creeks and rivers we have drunk from will fill with metals and turn orange? Will the fish survive this and the warming of the rivers and oceans? Will the caribou herds continue to decline, or will they adapt? Will the peoples who have dwelled in this land for millennia find a way to keep their culture alive?

Just how much of this place that I have come to love will change beyond recognition?

The answers are all a long way off. But my strong suspicion—and it is, at bottom, only that—is that if we truly grasped what hangs in the balance, if we sensed its magnitude in a way that even now I cannot, we would be doing much, much more. If we were a species of wiser, stronger, more capacious beings—ones who felt this planetary shock we are causing with the same vividness that one feels a charging bear—then we would find ways to price carbon. We would eat less meat. We would throw our weight behind transitioning to solar and wind and nuclear energy, which at this moment in history has begun to make a great deal of economic sense anyway. We would preserve unbroken tracts of land like the Brooks Range, which give ecosystems room to shift and adapt, and we would build more green spaces in urban areas, so that our children grow up sensitive to the wonders of the more-than-human world. Because if we do not care about the extinction of our fellow creatures—who are already disappearing by the hundreds each decade, at a rate perhaps faster than any time since an asteroid killed the dinosaurs—then what, beyond ourselves, could we possibly care about?

And yet, if we could somehow inhabit the wisdom of these hypothetical higher beings, I do not think we would panic. We would not despair. We would not retreat into ironic solipsism, or wallow in embittered self-pity, or give ourselves to apocalyptic trauma porn. We would understand that this planet was a wild and painful place long before we started burning fossil fuels, and will be so long after our kind has died out. Because we would see, finally, that life has always been hard. Its most basic facts have not really changed. We have never had much more than a toehold on this planet; have always lost what we loved most, and told ourselves stories to get along, and had far less control over any of it than we liked to believe. And then we have always died, and others have always replaced us, and the world has always spun on.

As higher beings, we would use our growing technological powers to see all of this a little more clearly. Not to absent ourselves, but to burn more brightly. We would practice observation, and nurture our attention. We would hold sacred the presence of mind to admire this world as it is, while we can.

I am not such a being. Such a being, I suspect, would not have spent much of the morning fantasizing about eating an entire pint of Ben & Jerry's at the first opportunity. But this land, and these people, have shown me precious things, which I will probably spend my whole life forgetting, and remembering, and trying not to forget again.

* * *

The final miles to Kivalina feel endless. A squall has kicked up over the coast, bringing rain that freezes our hands and winds that push us backward on the sluggish current. By the time we reach the mouth of the Wulik and begin rowing across the lagoon—the saltwater blowing in our eyes—the sun sits low behind gray clouds.

On the new bridge behind the village stands a man, waving walrus tusks at us; he reappears at the water's edge, directs us to a patch of land between the houses and the graveyard, helps us haul out our boats. Soon a troop of early adolescent boys appears. One points to the wad of tobacco in his mouth and tells us that his favorite rappers are Future and XXXTentacion. Another, age twelve, asks my favorite food and answers before I can: spaghetti. No, beluga muktuk. How many ex-girlfriends do I have? He has four.

We set up our tents on the lagoon side of the island, in the lee of a trailer, and then walk over to the sea. Battered by wind, its surface looks like a sheet of hammered metal. A century ago, during the Russian Revolution, a Moscow man named Tikhon Semushkin arrived at a similar lookout, having traveled to Chukotka to become a Bolshevik missionary. What he saw, when he arrived at the Bering Strait, was not the outpouring of life, or the strange beauty of the tundra, or the dignity of the people who had lived there for some ten thousand years. Instead, he announced the collision of "two days—New and Old—and two worlds, new and old, socialist and capitalist."

It is so easy to cover the world in notions, and never let it reach out and touch us. And for all of the miles I have walked and rafted and flown through this land, straining to bend my senses to it, I have traced only a thread. Were I to travel this same route in the opposite direction, or in a different season, or in a different stage of life, I would find something else. And this is only the outer crust: Beneath the surface churn ancient, future worlds still waiting for their day in the sun.

We snap a few pictures to remember this group of people, this moment of provisional completion. "One day we'll all come back here," says Maddy to Julia and Russell and me. "Like David Cooper." Then we split up to explore the village.

I follow the shoreline, watching waves peel along a sandbar, wondering if they might be surfable under the right conditions. Then I wonder what my friends are up to back in LA. *Who's home? Any concerts coming up?* I begin thinking about jobs and how I will support myself while turning these notes into a book. And as my thoughts race ahead, I stop walking, turn to the water, and try to let this place in one more time.

EPILOGUE

Not long after I returned to Los Angeles, a large mountain lion came down out of the Santa Monica Mountains into the city. Mountain lions do this sometimes when food is in short supply, which tends to happen during periods of drought. Southern California was then in the midst of a deep drought. So it was fitting that the lion was first sighted lounging on the Astroturf lawn of a Department of Water and Power building. He perked up by the time Fish and Wildlife arrived, and fled for several hours around multimillion-dollar homes and a conspicuously well-watered golf course before being tranquilized and released back into the mountains.

Growing up, I was vaguely aware that there were lions nearby, but I had never grasped the ecological significance of this fact. Lions are apex predators; their very existence implies the basic integrity of the entire ecosystem. Somewhere between ten and fifteen still live in the Santa Monica Mountains, which run right through the city. Their lives are hard. Isolated by the road system, they're often killed by cars or rat poison, and their populations are a good deal less genetically diverse than biologists would hope. But they are still alive, finding their way as we find ours.

What struck me about this particular encounter, aside from the fact that it happened along one of my favorite running routes, was how long it took us to spot the lion. Not until he had wandered two miles from the mountains, across Sunset and San Vicente Boulevards (two of the busiest streets in West LA), did the authorities receive a report—and by then it was 10:09 a.m. on a Thursday and the animal was across the street from an elementary school in session. Somehow one of the planet's largest land predators had wandered into our children's midst without anyone noticing.

The incident seemed emblematic of an ongoing attentional shift. Twelve months before, while I was still in Alaska, staying with Julia at Klara Maisch's home outside of Fairbanks, Facebook announced that it was changing its name to Meta and building a "metaverse"—a digital virtual reality in which users could work, socialize, recreate, and ostensibly live. "You can teleport…to wherever you want," promised founder and CEO Mark Zuckerberg in a promotional video. "Even just make things happen by thinking about them."

The announcement came just weeks after Facebook employee Frances Haugen leaked internal documents revealing that the company was fully aware of the platform's harmful effects on adolescents (the spikes in anxiety, eating disorders, self-harm, etc.), and had nonetheless prioritized profits over the well-being of its users. The incident did not inspire confidence in the corporation's fitness to midwife the future. What struck me most about the video, though, was how *flat* Zuckerberg's metaverse seemed. He'd reduced the entire more-than-human world to a child's fantasy, a grab bag of pleasing images.

"It has an incredibly inspiring view of whatever you find most beautiful," said a smiling Zuckerberg, standing before a virtual background of snow-dusted coniferous forest falling away into a palm-tree-adorned, apparently tropical coastline. Later he teleports into a "forest room" in which fish swim through the air. "Koi fish that fly?" observes a smiling robot. "That's new!"

I don't doubt that there will be fun and useful applications for virtual reality technologies, or that a well-grounded person will manage to take them in stride. But as a way of life, it seemed something like the opposite of "inspiring," to trade reality for a malleable image. To silo our minds off from a changing world even as we remain, inevitably, a part of it—eating, shitting, consuming energy. Zuckerberg's new technology seemed aimed at a part of us his earlier technologies had nurtured: the part that is anxious about the future, and thinks the world is dying, and just wants to check out. The part that's willing to trade the richness of connection for the illusion of control, and hasn't learned fidelity to anything bigger than ourselves.

Returning from Alaska, I still didn't know what the future held for me, but I knew I wanted to be someone who would notice the mountain lion.

* * *

A few months after the lion came down out of the mountains, a record-breaking series of storms pummeled Southern California, triggering mudslides in the hills, flooding the lowlands, and ending the drought. The following summer, for my twenty-sixth birthday, Julia drove down from Monterey—where she'd been doing a graduate degree in scientific illustration—to join me on a five-day hike through the Santa Monica Mountains.

We took the Backbone Trail, a recently completed sixty-seven-mile route that begins an hour's drive north of LA and ends in town, a few miles from my childhood home. I had run the northern end of the trail once in high school, several months after a fire swept through the region, and had seen hardly a single living thing; heat rose in shimmering waves from the charred ground. A decade later it was unrecognizable—a lush scrubland of sage and buckwheat and manzanita. Songbirds flitted between stands of laurel sumac draped with wild cucumber; red and purple and yellow wildflowers popped along the trail. A baby hummingbird fluffed itself on a twig, accepting a droplet of regurgitated nectar and bugs from its mother.

Deeper in the mountains, the landscape changed and changed again. We walked through grassy meadows and arid chaparral and tangled woodlands of live oak and hollyleaf buckthorn and poison oak. We followed creeks that flowed through sycamore-lined valleys, and descended shady hillsides of bay laurel. Along a barren ridge, we found a charred, skull-sized cone sticking out of a pile of landslide rubble—evidence, perhaps, of a wetter time when gray pines grew up high. Everywhere there were creatures: Coveys of quails, their plumes quivering ridiculously as they fled from us. A gray fox, swishing its tail on a log. A red-tailed hawk chasing green parrots, and rattlesnakes coiled on sun-baked stone.

Julia had spent her free time in Monterey learning the ecology of the Central Coast. "I'd thought it was all buildings and beach," she admitted. For most of my life, so had I. But soon she'd recognized species from Alaska—white-crowned sparrows, willows, lupine—and realized she could apply the same ecological attention here.

Down in LA I'd been doing the same, spending some time outside each day before beginning my afternoon job as a high school tutor and my night job as a journalist. Slowly I'd been learning the seasonal patterns of the plants and the calls of the birds, and now we pooled our knowledge, using our phones to learn more. An AI application called Plantum told us that a large, alien-looking succulent just off-trail was called *Dudleya pulverulenta*, and that its three-foot-long, neon-red flowers had probably evolved to attract hummingbird pollination. The Merlin Bird ID app listened to a mockingbird with us and found that it imitated the calls of twelve other bird species in as many minutes.

On the afternoon of my birthday, I told Julia something Roman had admitted in a recent text exchange: His bad hip had not really forced him to leave our first expedition early. The pain was bad, but he'd pushed through worse before. He'd left because he thought we'd learn more on our own.

"He was sowing his own seeds," observed Julia.

Indeed he was. At that moment, Russell and Maddy were gearing up for another trip with him to the Brooks Range, this time to study the outbreak of orange rivers. Julia was about to return to Alaska to begin making graphics for the National Oceanic and Atmospheric Administration, and I was planning on starting a PhD as soon as I finished this book, to do more research in the Arctic. There was so much to do. Even down here, signs of change were all around. It was unseasonably warm as we hiked, and the mountains were pocked by mudslides from the winter. We knew the droughts and fires would return. (A few years later, the worst fires in LA's history would burn thousands of homes and force my parents to evacuate.) But this trip, Julia and I had agreed, was for watching.

"It's so beautiful," she said one night, as we lay outside on our sleeping pads, watching the Big Dipper scoop water out of the Pacific.

On our last evening, as we walked through tunnels of chamise and California lilac shrubs, I paused to identify a bird: a wrentit scrounging for insects in the end-of-day light. I started walking again, but Julia held still. "Listen," she said.

At first I didn't know what she was talking about. Then I heard it: a percussion of nano-sized *pop*s. Dozens every second; a constant flurry softer than whispers. There were little flashes of movement under the shrubs, and at the same moment, Julia and I both realized: The nut-like fruits of the bigpod lilacs were bursting open in the July warmth, expelling seeds with force.

These seeds, we'd learn the next day—after traipsing through town to my parents' home for showers and dinner, and then doing a Google search—tend to lie dormant for years, centuries even, until a wildfire heats them open. Then they germinate, beginning anew. We didn't know this at the time. What we knew was that bigpod lilacs are among the most common plants in these mountains. We'd been walking through them all afternoon. I'd been walking through them my whole life.

ACKNOWLEDGMENTS

"Well, do you know what you're doing?"

Matt Nolan asked me this in the spring of 2019, when I first proposed following him to McCall Glacier to write about his work. It was a fair question. At the time, I had published only a handful of articles for small local papers, and I could provide no credible assurance that my project would go anywhere. But when I mentioned this exchange during a conversation with John McPhee, John cut in: "Give him my number. Tell him he can call me for a reference." Matt—who, like many in Alaska, admired John's work immensely—never called. John's gesture was enough.

It is no bit of faux modesty to say that this book might never have happened without John's inspiration, instruction, and mentorship. But several others played a similarly vital role: Rob Nixon, who first advised my Princeton undergraduate thesis on climate change in Alaska and later encouraged me to turn it into a book. My agent, Bonnie Nadell, who not only pulled this project from her slush pile, but spent more than a year helping me figure out what I was trying to say. My editor,

ACKNOWLEDGMENTS

Karyn Marcus, who believed in this project from the beginning and read countless half-baked drafts along the way. Karyn's assistant, Ian Dorset, who showed editorial judgment well beyond his years.

My brilliant mentor and friend Kristin Dombek also played a critical role in developing my narrative voice, refining my thinking on technology, and curbing my excesses. Zahid Chaudhary, Rebecca Mead, Heller McAlpin, Meera Subramanian, and Erika Milam all mentored me in my writing and/or scholarship. Special thanks to Jessica Woollard for her early input. And thanks to my sister, Amy Weissenbach, and the many friends who provided critical feedback throughout: Austin Coffey, Coby Goldberg, J.J. Woronoff, Julia Ditto, Russell Wong, Heather Nelson, Lydia Weintraub, Joshua Judd Porter, Ryland Marcus, and others.

I am grateful to the many unmentioned or unquoted experts and locals who answered my questions—a too-long list that includes Matt Sturm, Rob Kaler, Kathy Kulutz, Gabriel Vecchi, Ricky Ashby, George Unalek, Vince Onalik, Chris Connors, Elie Gurarie, Jack Hébert, David Lawrence, Ned Rozell, Jon Holmgren, Patrick "Pat" Burns, Elena Shevliakova, Pat Holloway, Lorene Lynn, and others.

Thanks to Jonathan, Jenny, Julia, and Laura Ditto; Tammy, Don, Hannah, and Savanna Bradley; Chris, Mary, Klara, and Jordi Maisch; Peggy Dial; Griffin Hagle; Mark Marette; Sue Aikens; James Smith; Mark Lockwood; Wil Gerken; Billie's Backpackers Hostel; and the people of Noatak Village for generously hosting me in the course of my travels.

Thank you to Princeton University (which provided a Martin A. Dale '53 award and an A. Scott Berg Fellowship), the Henry Luce Foundation, and the St. Anthony Hall Educational Foundation for funding my field research.

ACKNOWLEDGMENTS

Thank you to my mother, Ann Southworth, for teaching me how to write. Thank you to my father, John Weissenbach, for teaching me how to edit.

Most importantly: Thank you to Matt Nolan, Kenji Yoshikawa, and Roman Dial for the adventure of a lifetime.

ABOUT THE AUTHOR

Ben Weissenbach is a writer from Los Angeles. He studied under John McPhee at Princeton University and was awarded a Gates Cambridge Scholarship to pursue a PhD in polar studies. His work has appeared in the *LA Times*, *National Geographic*, *Scientific American*, and *Smithsonian*, among other publications.